Approaching Home Automation

A Guide to Using X-10 Technology

Bill Berner - Craig Elliott

Approaching, Inc.
Chicago • San Francisco

Library of Congress Cataloging in Publication Data

Approaching Home Automation: A Guide to Using X-10 Technology /
    Bill Berner...Craig Elliott
p. cm.
ISBN 1-881911-00-4 (soft cover)
1. Electronics

Apple, the Apple logo, and Macintosh are registered trademarks of Apple Computer, Inc.

IBM is a registered trademark of International Business Machines Corporation.

MacPaint is a registered trademark of Claris Corporation.

POWERFLASH, POWERHOUSE, PROTECTOR PLUS, and X-10 are trademarks of X-10 (USA), Inc.

Library of Congress Catalog Card Number 92–97022

Copyright ©1993 by Approaching, Inc.

All rights reserved. No part of this publication may be reproduced, stored in a retrieval system, or transmitted, in any form or by any means, electronic, mechanical, photocopying, recording, or otherwise, without the prior written permission of the publisher.

Although all possible measures have been taken to ensure the accuracy of the material presented, neither the author nor Approaching, Inc. is liable in case of misinterpretation of directions, misapplication, or typographical error.

Cover design by Dale Engelbert

Printed in the United States of America

Preface

The advent of electronics has brought us many wonderful technologies. Calculators, digital watches, and personal computers are taken for granted in our everyday lives. As these silicon marvels came into being, one of the applications thought to be imminent was the "automated home." The idea was to use electronics to control lights and appliances, monitor security, and make life in general more comfortable. In fact, Steve Wozniak's open design of the Apple II computer was largely to facilitate the development of home-control devices. Woz was confident that the major application for his new personal computer would be home automation.

It never really happened.

Sure, you can buy an electric iron that shuts off if it hasn't been moved for a period of time, and if you watch enough late-night TV you can purchase devices that will turn lights on and off at the sound of clapping hands, but the dream of home automation is largely unrealized.

The big question is, "Why?" Simple, inexpensive, and readily available, the X-10 system has been on the market since 1978 and provides an excellent means to implement many home automation tasks. Other technologies have promised marvelous cutting-edge features that will surpass X-10. Until that happens (and don't hold your breath), X-10 provides a wonderful mechanism to explore home automation inexpensively—today.

This technology was invented in the late 1970s by a company called X-10 and distributed by BSR, Inc. Just as important was X-10's licensing of technology to literally hundreds of other electronics manufacturers for inclusion in their products. The combination of products from the X-10 company and others makes this the most widely used home-automation system in the world.

This book is designed to give you an understanding of what X-10 is all about. It describes the components of an X-10 system and provides design guidelines to create a useful, functional automated home. Because of the huge number of X-10–compatible products, we can cover only the most common ones. They are all available from X-10 (USA), Inc. and its distributors. You'll find other products with similar (if not identical) features from other manufacturers. Most of the information applies to those products, but check the product manuals for differences.

We've also included a technical overview, a troubleshooting section, a list of X-10-compatible manufacturers, and other information we thought might be helpful.

We also give safety guidelines when discussing electrical equipment. All the products described in this book are listed with the Underwriters Laboratories for product quality and safety, but you still need to use common sense. The X-10 products in this book require at most the same electrical skills as are required for installing a dimmer control. But that still means dealing with potentially lethal electricity. Contact a qualified electrician if you have any questions or doubts. It's a relatively inexpensive way to gain peace of mind.

We'd like to thank Dave Rye, Vice President and Technical Manager of X-10 (USA) Inc. Dave is an expert in the field of home automation, with over 15 years of experience, and has written many articles on the subject. His expert advice and review added unique value to this book. We're convinced he knows everything.

We would also like to thank Jim Berner (two of them), Dave Lenkaitis, Dan Torres, Ernie Zamora, Teri Thomas, and Bill Scheffler for their efforts in reviewing, editing, and generally keeping us honest.

We were amazed at the lack of information that was available about X-10 and its applications. It's not difficult, just not very well known.

Most of all, we think this stuff is fun. That's why we wrote this book.

We hope you like it.

Contents

1 Concepts 1

Overview 1
Basics 2
Controllers 3
Modules 4
Compatibility 7

2 Controllers 9

Overview 9
Maxi Controller 10
Mini Controller 14
Mini Timer 17
Using the Mini Timer for Timed Events 18
Home Automation Interface 27
Telephone Transponder 28
Remote Control and Wireless Transceiver 33

3 Controlling Lights 37

Overview 37
Lamp Module 38
Two-Way Wall Switch Module 40
Three-Way Wall Switch Module 42
Installing a Three-Way Wall Switch 43
Applications 45

4 Controlling Appliances 47

Overview 47
Modules 48
Two- and Three-Pin Appliance Modules 48
Wall Receptacle Module 50
Thermostat Set-back Controller 52
Heavy-Duty Appliance Module 55
Universal Module 57

5 Home Security 67

Overview 67
Home Security Basics 68
Powerflash Burglar Alarm Interface 70
Supervised Home Security System 74
Home Security Options 85
Using the Supervised Security System 89

6 Apple Macintosh 93

Overview 93
Setup 94
Getting Started 96
Testing the Interface 99
Creating and Using a Module Icon 100
Setting Up Timed Events 106
Setting the Base Housecode 118
Configuring and Customizing 121
Instant X-10 129
Summary 132

7 IBM PC 133

Overview 133
Setup 134
Entering Module Information 139
Erasing Module Information 143
Controlling Modules 144
Setting Timed Events 148
Saving, Printing, Exiting, and Other Activities 153

A Technical Overview 157

Overview 157
How It Works 157

B Troubleshooting 163

Overview 163
Things Don't Work at All 163
Things Don't Work Right 165

C Compatible Products 167

Glossary 175

Index 181

1 Concepts

Overview

This chapter explains the concepts you'll need to understand before you begin automating your house. Because we know you're eager to get started, we've kept this introductory chapter as short as possible, but it does contain important information, and you should read it carefully before continuing.

First you'll learn about the basic components of a home-automation system: **controllers** and **modules.** Next you'll learn how controllers communicate with modules by sending **commands** to them. You'll learn about the specific commands you can use, and how controllers can send commands to particular modules by using the module's **address.**

When you've finished this chapter, you should have a solid understanding of **X-10 Home Automation** components and their operation, and you should be ready to learn about the different kinds of controllers and modules you can use to build your system.

Basics

You need two types of devices to automate your home: controllers and modules. Modules are adaptors that you connect to light switches, lamps, appliances, or other devices that you want to control. Controllers send commands to modules to control the device attached to the module.

The diagram below shows the simplest X-10 configuration: a single controller sending commands to a single module. The controller shown here is used to turn the lamp on and off, and to dim or brighten it.

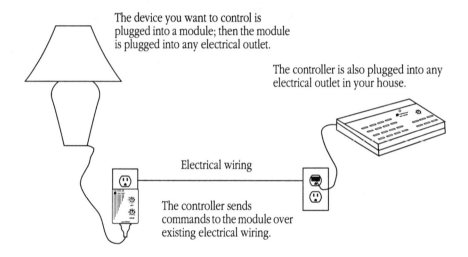

The device you want to control is plugged into a module; then the module is plugged into any electrical outlet.

The controller is also plugged into any electrical outlet in your house.

Electrical wiring

The controller sends commands to the module over existing electrical wiring.

You can buy modules to control just about anything: lights, appliances, sprinklers, security systems—even draperies. In general, any controller can be used to control any of the modules, and we will often use the term **device** to refer generically to something that can be attached to an X-10 module. A complete description of the most popular controllers is given in Chapter 2, "Controllers," on page 9. Modules you can use to control lights are in Chapter 3, "Controlling Lights," on page 37, and modules that control appliances are described in Chapter 4, "Controlling Appliances," on page 47.

You may be asking, "Why not just turn the lamp on and off by hand?" That's a good question. In fact, there will be times when you'll want to turn the light on and off by hand, and you'll still be able to do that once you're using X-10 controllers and modules. But if you've ever stumbled through a dark house to find a light switch, or left a garage light on all night (or all week), or been startled by a noise in the middle of the night, or gone on vacation and worried about your house looking empty, then an X-10 system can be an inexpensive way to make your life a whole lot easier.

At this point, you're probably thinking, "Well, that's all great, but I'll bet it's a pain in the neck to set up the system." Wrong. X-10 controllers and modules communicate over the existing electrical wiring in your house, so you don't need to install new wiring. You just plug modules and controllers into existing electrical outlets. The X-10 devices can send commands over the same wire that carries electricity throughout the house (without disrupting your electrical service in any way, of course).

The technology behind all this is a little complex, and you certainly don't have to understand it to use X-10 products. But if you're curious, see Appendix A, "Technical Overview," on page 157.

Controllers

You've already learned that there are two components to every X-10 system: controllers and modules. You also know that controllers send commands to modules and that modules execute those commands to control the devices attached to them.

When you want to control a device, you push one or more buttons on the face of the controller. (Or, if you are using the Home Automation Interface, you enter commands at your computer.) When you push a button, the controller sends a command to the module or modules that you've selected, and the modules control the devices attached to them.

This section describes the commands that controllers use to communicate with modules. There are two types of commands: **address commands** and **function commands.** Address commands identify the modules you want to control. Function commands tell the modules what to do.

Address Commands

Each X-10 module has a specific address, which you'll learn more about in the next section. X-10 controllers use these addresses to send function commands to a particular module. This is how an X-10 controller can turn on a light in the living room without turning on every other device in the house at the same time.

When the controller sends an address command, the module or modules with that address "wake up" and begin listening for a function command. As soon as they "hear" the function command, they perform that function.

Function Commands

Here's a description of the different X-10 function commands:

ON/OFF You can use X-10 controllers and modules to turn lights, appliances, and other devices on and off.

DIM/BRIGHT You can also use X-10 controllers to dim incandescent lights—even lights without a dimmer control.

ALL LIGHTS ON Some controllers allow you to turn on all of the lights in the house with a single button. This is especially useful in an emergency—for example, if you suspect that someone is breaking in. With ALL LIGHTS ON, you can turn on every light in the system at once, possibly scaring away the intruder.

ALL UNITS OFF Controllers that support ALL LIGHTS ON also support ALL UNITS OFF. This turns off all the devices attached to modules with the same Housecode (more about Housecodes later). ALL UNITS OFF provides a convenient way to turn off all the lights in the house after you're in bed, or to make sure that you've turned off all your appliances before leaving town.

Modules

Modules are adaptors that you install between the device you want to control and the source of electricity for the device. For example, the Lamp Module is a small interface box that you plug a lamp into. Then you plug the Lamp Module into an electrical outlet. Because the Lamp Module sits between the source of electricity (the outlet) and the device (the lamp), it can control the device by regulating the amount of electricity the device receives.

Other modules replace standard light switches and control the lights that the light switches turn on and off. Still others can be used to control low-voltage electrical devices, such as lawn sprinklers and outdoor lighting.

In the previous section, we introduced the concept of the module address, which controllers use to send function commands to specific modules or groups of modules.

X-10 addresses are similar to the addresses that the post office uses to deliver mail. The post office can identify an individual house because each house has a unique address: a street number and name, a city, a state, and a zip code. X-10 addresses are a little simpler than postal addresses because they have only two components, a **Housecode** and a **Unit Code.**

You set the Housecode and Unit Code of a module by adjusting two dials on the front of the module. Typical dials are shown below.

HOUSE CODE   UNIT CODE

Each dial has 16 different settings. The Housecode is selected from the letters A through P, and the Unit Code is selected from the numbers 1 through 16. The address of the module is simply its Housecode followed by its Unit Code. So, for example, the address of the module shown above is A1.

Because there are 16 different Housecodes and 16 different Unit Codes, there are 16 x 16 = 256 possible addresses. That should be more than enough for the average house. Also, you can assign two or more modules the same address if you want to control them simultaneously. For example, you might have two porch lights that you want to go on and off at the same times. In that case, you can assign both modules the same address.

In general, you can assign addresses to modules however you want, but there are some guidelines you might want to follow to make your system easier to use. We'll give a few suggestions here.

- Select devices that you *always* want to control at the same time and give them the same address (the same Unit Code and Housecode).

By assigning the same address to all the modules in a group, you will be able to turn all of them on or off simply by pushing a single button on the controller. In addition, you'll be able to create **timed events** to control all of these modules with either a Mini Timer or a Home Automation Interface (see Chapter 2, "Controllers").

- Select devices that you will *occasionally* want to control simultaneously and give them the same Housecode.

This will allow you to control the group of devices with the ALL LIGHTS ON and ALL UNITS OFF commands. For example, you might have ten different lights in your home. By giving them all the same Housecode, you can turn all of them off at night from a single controller on your nightstand. But at the same time, you might have other devices—for example, a stereo or an outdoor porch light—that you *don't* want to turn off when you go to bed. In that case, you'd assign these lights and appliances a *different* Housecode from that of the rest of the lights, so that the ALL UNITS OFF command would not turn them off.

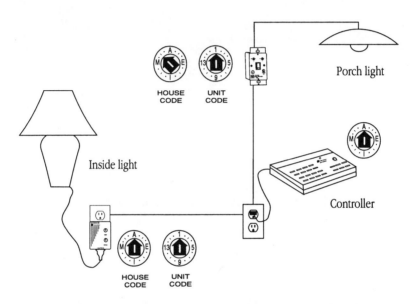

In the diagram above, the ALL UNITS OFF button on the controller will turn off the inside light, but not the porch light. That's because the porch light has a different Housecode.

Also, the buttons on a controller will send commands to only one Housecode. You must adjust a dial on the controller to send commands to modules with a different Housecode. By grouping modules with similar functions, you can control the entire group with a single controller. For example, if you have five sprinkler heads in the yard and ten lights in the house, you could have one controller next to your bed for controlling all the lights, and another on the porch for controlling all the sprinklers.

Compatibility

The set of commands described in this chapter, and the method for transmitting them over standard electrical wire, were established in the late 1970s by a company named X-10 (USA), Inc. Literally thousands of X-10–compatible home-automation products are in use today.

X-10–compatible products are available from a variety of vendors, some of whom are listed in Appendix C. Generally, because these products all use the same, standard set of commands, you can mix and match products from different vendors in the same system. For example, you can use a Radio Shack controller with an X-10 Lamp Module.

A word of warning: Although two products from different vendors may do exactly the same thing, their names are liable to be quite different. Occasionally, you might have trouble deciding whether you're buying the right thing. In this book, we use the names established by X-10 (USA), Inc. If you are not sure whether a similar product is compatible, you may be able to find out by asking a knowledgeable salesperson. If that doesn't work, try calling the company. Phone numbers are also listed in Appendix C.

2 Controllers

Overview

Chapter 1 explained the basic concepts of X-10 home automation: Controllers send commands to modules, and the modules execute the commands, turning devices on or off, dimming lamps, and so on. This chapter describes the different kinds of controllers you can use for your X-10 system.

Each section talks about a different controller. The section begins with a drawing of the controller and an explanation of its major components. Once you've become familiar with X-10 technology and products, you should be able to answer most of your controller-related questions just by looking at these diagrams. Later in the section, we list the important features that distinguish one controller from another, and that make some controllers more appropriate for a particular application. We also provide some suggestions for using each controller in your home.

Once you've read this chapter, you should be able to select the controller or controllers that are most appropriate for your system and your plans. You should also know which controllers you can use to add features to your system in the future.

Maxi Controller

Overview

The Maxi Controller is a general-purpose controller that can be used to control up to 16 devices. Here's a diagram of the Maxi Controller with a description of the major components.

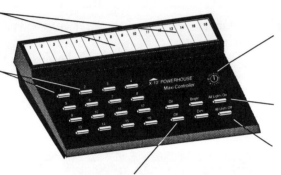

Write the names of devices you're controlling at the top of the console for quick reference.

Click one or more of the 16 **Unit Code buttons** to identify the Unit Codes of the modules you want to control.

After selecting one or more devices, press ON, OFF, BRIGHT, or DIM to control the selected device or devices.

Set the Housecode here. Make sure it matches the setting on the modules you want to control.

Press ALL LIGHTS ON to turn on all lights.

Press ALL UNITS OFF to turn off all modules with the same Housecode as the Maxi Controller, including Appliance Modules.

Using the Maxi Controller

Using the Maxi Controller is easy. Here's what you do:

Plug the Maxi Controller into any electrical outlet.

Don't forget this step, or the Maxi Controller won't be able to send commands, and nothing will work.

Set the Housecode on the Maxi Controller to the same setting as the Housecode on the module or modules that you want to control.

With the Maxi Controller, or any other controller, you can send commands *only* to modules that have the same Housecode as the controller. This means you can control only 16 devices (with distinct Unit Codes) at a time. You can, of course, control more than 16 devices by assigning the same Unit Code to more than one module.

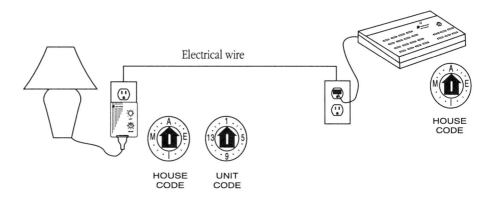

The Housecode on the controller *must* match the Housecode on the module. A controller set to Housecode A, like the one shown above, will not turn on a module with a Housecode other than "A." For example, it would not turn on the module shown below, which is set to Housecode O. For that module, you would have to use a controller set to Housecode O, like the one shown below.

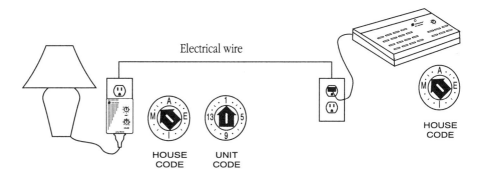

You can change the Housecode by inserting a small screwdriver in the dial in the upper-right corner of the controller and turning to the desired Housecode.

Press one or more Unit Code buttons to send address commands to the modules you want to control.

You select the modules that you want to control by pressing the appropriate Unit Code buttons. For example, if you want to turn off a lamp that is connected to a module with the Unit Code set to 10, then press the button marked "10." When you do this, the controller sends an address command that consists of the Housecode you've set on the Maxi Controller plus the Unit Code corresponding to the button you press. This causes any modules with that address to "listen" for the next function command.

Note that you can press more than one Unit Code button. For example, if you want to turn off two space heaters that have Unit Codes 3 and 4, you'd press the button marked "3" and then press the button marked "4." When the first module sees the first address command with Unit Code 3, it will wake up, and it will ignore the next address command, with Unit Code 4, while continuing to listen for a function command. By pressing multiple Unit Code buttons before pressing a function command button, you can control groups of modules, which is an important feature of the Maxi Controller.

Next, choose the function commands that you want to send to the module(s).

You send function commands to the selected modules by pressing the ON, OFF, DIM, and BRIGHT buttons. Note that you can dim or brighten only lamps that are connected to Lamp Modules or Wall Switch Modules.

That's it! That's all you need to know to control individual devices or groups of devices with the Maxi Controller.

ALL LIGHTS ON and ALL UNITS OFF

There are times when you might want to turn on all of the lights with the same Housecode at once, or turn off all devices with the same Housecode at once. Here's how you use these additional features:

Make sure that the Housecode on the Maxi Controller is the same as the Housecode on the modules you want to control.

You can change the Housecode by inserting a small screwdriver in the dial in the upper-right corner of the controller and turning to the desired Housecode.

Press ALL LIGHTS ON or ALL UNITS OFF.

Press ALL LIGHTS ON to turn on all the lights connected to Lamp Modules and Wall Switch Modules that have the same Housecode as the Maxi Controller.

ALL UNITS OFF works like ALL LIGHTS ON, but it turns off all the modules that have the same Housecode as the Maxi Controller, including Appliance Modules.

It's important to remember that controllers can control only modules that use the same Housecode that the controller uses.

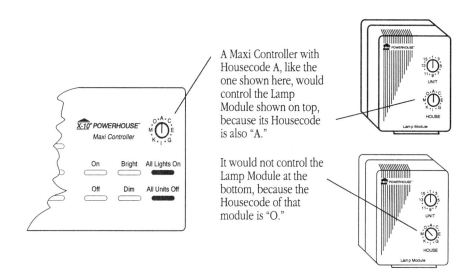

A Maxi Controller with Housecode A, like the one shown here, would control the Lamp Module shown on top, because its Housecode is also "A."

It would not control the Lamp Module at the bottom, because the Housecode of that module is "O."

Comparing the Maxi Controller with Other Controllers

The Maxi Controller is perfect when you are home and able to turn appliances, lights, and other devices on and off. For example, it lets you turn lights off from your bedside before going to sleep, or turn off basement or garage lights from inside the house.

The Maxi Controller cannot be used for timed events. For example, you can't set the Maxi Controller to turn sprinklers on and off at a specified time every day. For this type of control, you need to purchase either a Mini Timer or a Home Automation Interface. The Home Automation Interface works in conjunction with a personal computer; the Mini Timer doesn't require a computer. Both controllers are described later in this chapter.

Mini Controller

Overview

The Mini Controller is a low-cost way to control up to eight modules. It can brighten or dim lights, and it can turn all lights on or all units off.

The Mini Controller is also ideal as a second controller used in conjunction with a Maxi Controller or a Home Automation Interface (described elsewhere in this chapter).

Here's a diagram of the Mini Controller with a description of the major components.

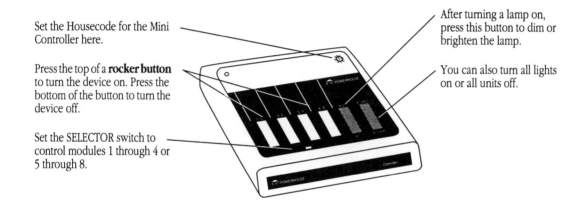

Set the Housecode for the Mini Controller here.

Press the top of a **rocker button** to turn the device on. Press the bottom of the button to turn the device off.

Set the SELECTOR switch to control modules 1 through 4 or 5 through 8.

After turning a lamp on, press this button to dim or brighten the lamp.

You can also turn all lights on or all units off.

Using the Mini Controller

To use the Mini Controller, here's what you do:

Plug the Mini Controller into any electrical outlet.

If you forget this step, you won't be able to communicate with anything.

Make sure that the Housecode on the Mini Controller is the same as the Housecode on the modules you want to control.

Like any other controller, the Mini Controller will communicate only with modules that use the same Housecode it uses. If the Housecodes are not the same, nothing will happen. If you have problems controlling a lamp, an appliance, or anything else, this is the first thing you should check.

Remember that you can change the Housecode by inserting a small screwdriver in the dial in the upper-right corner of the controller and turning to the desired Housecode.

Determine which module you want to control, and set the SELECTOR switch appropriately.

The Mini Controller works differently than the Maxi Controller does. Remember that the Maxi Controller has separate Unit Code buttons and function command buttons. The Mini Controller has combined Unit Code and ON/OFF buttons—called "rocker buttons" because they rock back and forth. If you press one side of the button, the controller sends an address command followed by ON; if you press the other side, the controller sends an address command followed by OFF. The ON and OFF sides are labeled clearly on the controller so you won't get confused.

Although the Mini Controller can control up to eight devices, it has only four rocker buttons. Each rocker button can control one of two Unit Codes, depending on how you have set the SELECTOR switch.

If you want to control a module with Unit Code 1 through 4, push the SELECTOR switch to the left. Push the SELECTOR switch to the right to control a module with Unit Code 5 through 8.

Press the appropriate rocker button to turn the module on or off. Or press ALL LIGHTS ON or ALL UNITS OFF.

Each of the white buttons on the front of the Mini Controller is labeled with two numbers corresponding to the Unit Codes of the modules that the button controls. For example, the first button is labeled "1•5." If the SELECTOR switch is set to the left, this button will turn the module with Unit Code 1 on or off. If the SELECTOR switch is set to the right, the button will turn the module with Unit Code 5 on or off.

Instead of pressing one of the white buttons to turn an individual module on or off, you can press ALL LIGHTS ON to turn on all the lights connected to Lamp Modules or Wall Switch Modules that have the same Housecode as the Mini Controller. Alternatively, you can press ALL UNITS OFF to turn off all devices connected to modules with the same Housecode as the Mini Controller.

Each rocker button on the Mini Controller turns modules both on and off. To turn a module on, press the top of the button. To turn a module off, press the bottom of the button. Remember that ALL LIGHTS ON and ALL UNITS OFF control only modules that have the same Housecode as the Mini Controller.

After you've turned on a light, press the BRIGHT/DIM button if you want to change the intensity of the light.

Remember that you can dim or brighten only lamps that you are controlling with either a Lamp Module or a Wall Switch Module. You can even dim or brighten lights that have been turned on using ALL LIGHTS ON.

Note that if you press more than one rocker button at the same time, you can turn lights on as a group, and then dim or brighten them as a group. For example, if you press rocker buttons 1 and 2 simultaneously, the Mini Controller will send "wake up" commands to modules 1 and 2, followed by ON. You can then change the intensity of both lamps by pressing DIM or BRIGHT.

That's it! That's really all you need to know to use the Mini Controller to control devices in your home.

Comparing the Mini Controller with Other Controllers

The Mini Controller is one of the least expensive controllers, and is ideal if you want to turn a small number of modules on and off. The Mini Controller doesn't support timed events, so if you need to turn devices on and off at specific times throughout the day, the Mini Timer or the Home Automation Interface would be a better choice.

The Mini Controller works well for controlling small groups of lights or appliances. Here are some examples.

- Turn a group of lights off from your nightstand after going to bed.
- Turn a group of lights on when you walk in the front door at night.
- Turn individual sprinklers on and off from inside the house.
- Turn your computer, printer, and other peripherals on and off from your desk.

Mini Timer

Overview

The Mini Timer combines the features of the Mini Controller and an alarm clock, allowing you to control modules both manually and using timed events. For example, you can use the Mini Timer to automatically turn outdoor lights on every evening and off every morning.

Here is a diagram of the Mini Timer, with a description of the major components.

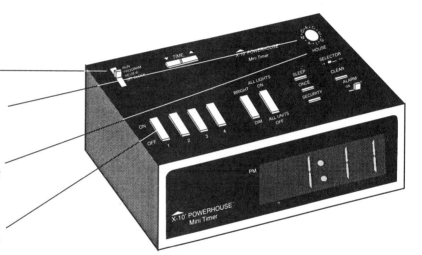

Use the MODE switch to prepare the Mini Timer for a particular operation.

Set the Housecode of the Mini Timer here.

Set the SELECTOR switch to control modules 1 through 4 or 5 through 8.

Press the top of a rocker button to turn the device on. Press the bottom to turn the device off.

Using the Mini Timer for Manual Control

In manual mode, you use the Mini Timer the same way that you use the Mini Controller. Here's an overview of the steps, in case you've forgotten:

- Choose a Housecode and set it using the dial at the upper right of the control panel.
- Set the SELECTOR switch to control modules 1 through 4 or 5 through 8.
- To turn a device on, press the top of the rocker button that corresponds to the Unit Code of the device you want to control; to turn a device off, press the bottom of the rocker button.
- Once you have turned a lamp on, you can brighten or dim it by pressing the top or the bottom of the BRIGHT/DIM rocker button.
- You can also use the ALL LIGHTS ON and ALL UNITS OFF buttons to control all modules with the same Housecode as the Mini Timer.

Remember that pressing a rocker button turns on or off *all* the devices that have the same Housecode as the Mini Timer and the same Unit Code as the rocker button. This allows you to control multiple lamps, appliances, or other devices at the same time.

If you are still not sure how to use the Mini Timer to control devices manually, review the steps outlined in the Mini Controller section. The steps for controlling modules with these two controllers are identical, and the steps are given in more detail in the Mini Controller section. The next section describes how to use the Mini Timer to create timed events.

Using the Mini Timer for Timed Events

Overview

You can also use the Mini Timer to create timed events, which gives you the power to automate your home in many useful ways. For example, you can create a timed event to turn your bedroom light on at 6:00 AM every morning to help you wake up.

Setting the Clock

Before doing anything else, you need to set the clock on the Mini Timer. Until you do this, your timed events won't happen at the right time!

Here's how to set the clock:

Plug the Mini Timer into a standard electrical outlet.

When you plug the Mini Timer in, the display flashes.

Next, you'll tell the Mini Timer that you want to set the clock. You do this by using the MODE switch, which is at the upper left of the Mini Timer controls.

Set the MODE switch to CLOCK SET.

Refer to the diagram at the beginning of this section if you have trouble locating the MODE switch.

Now you're ready to set the time.

Set the time by pressing the rocker button marked "TIME."

To advance the time, press the side of the rocker button marked "▲." To move the time back, press the side of the button marked "▼." Note that pressing the button and releasing it quickly moves the time slowly; pressing and holding the button moves the time rapidly. Also note that the PM indicator light comes on for PM times and turns off for AM times. Be sure that you have correctly set the time for AM or PM. This will be very important when you set the Mini Timer to turn devices on automatically—you wouldn't want it to brew the coffee at 6:00 PM or turn the porch light on at 6:00 AM.

Now that you've set the time correctly, you need to change the mode of the Mini Timer so it's ready to work.

Set the MODE switch to RUN.

Setting Timed Events

Now the clock is set, and you've learned to use the Mini Timer to control modules manually. The real power of the Mini Timer, though, is its ability to use timed events to control modules automatically at different times throughout the day.

The procedures outlined here will allow you to turn a module (or group of modules with the same Unit Code) on or off automatically. For example, you could turn the porch light on at 6:00 PM every day.

You can create timed events for only one group of four modules. You can choose from modules with Unit Codes 1 through 4 or modules with Unit Codes 5 through 8. You choose which group you want to control by using the SELECTOR switch on the face of the Mini Timer.

In other words, you cannot set timed events for both sets of Unit Codes at the same time. For example, if the Housecode on the Mini Timer is "A," you can create timed events for modules with addresses A1, A2, A3, and A4 *or* modules A5, A6, A7, and A8. But you cannot simultaneously control modules from both groups.

In fact, if you program timed events for the modules in one group, and then change the SELECTOR switch setting, the events you've programmed will occur for the modules in the other group! For example, let's say you control a porch light with a module that uses Unit Code 1, and a space heater with a module that uses Unit Code 5. If you set the SELECTOR switch to 1-4 and program an event to turn on the porch light at 6:00 AM every day, and then move the SELECTOR switch to 5-8, the space heater will come on at 6:00 AM every day instead of the porch light. This kind of mistake could be dangerous, and you should be sure you understand how the Mini Timer works before you use it to create timed events.

You should also know that you can create only two "on times" and two "off times" for each of the four Unit Codes in the group you've chosen. While this may be enough for many home-automation systems, you will need to use a Home Automation Interface if you want to program more than two on times or two off times.

To create timed events, you first put the Mini Timer into PROG SET mode, which is short for "Program Set" mode.

Set the MODE switch to PROG SET.

When the Mini Timer is in Program Set mode, the display shows you the events that have already been programmed. If you have not programmed any events, the display will show "12:00."

Next, you need to set the time that you want the event to occur.

Use the TIME button to set the time of the event. Pressing ▼ moves the time back. Pressing ▲ moves the time forward.

Note that the PM indicator turns on for PM times and turns off for AM times. Be sure that you have correctly set the time for AM or PM.

You must also set the SELECTOR switch appropriately for the module or modules that you want to control.

Set the SELECTOR switch to 1-4 or 5-8, depending on the Unit Code of the module you want to control.

Note that if you set the SELECTOR switch to 1-4, create timed events, and then change the SELECTOR switch to 5-8, events that you created for modules with Unit Code 1 will occur for modules with Unit Code 5, events you created for modules with Unit Code 2 will occur for modules with Unit Code 6, and so on.

For example, let's say you control a lamp with a module set to Unit Code 2, and a space heater with a module set to Unit Code 6. If you set the SELECTOR switch to 1-4 and program events for the lamp, and then change the SELECTOR switch to 5-8, all of the events you programmed for the lamp will happen for the heater.

Next, you'll program the event.

Press ON or OFF on the rocker button that corresponds to the Unit Code of the module you want to control.

For example, if you want to turn on the module with Unit Code 1, first set the SELECTOR switch to 1-4, and then press ON for the rocker button labeled 1•5.

Now you can program additional events by following these steps:

- Set the time for the event.
- Press ON or OFF for the Unit Code of the module you want to control.

If "18:88" appears on the display, the memory for the selected Unit Code is full. You must clear one of the currently stored events before setting a new one. See "Reviewing Timed Events" on page 24.

Once you have finished setting events, you'll need to reset the MODE switch for normal operation. If you do not, the Mini Timer won't execute any timed events.

When you have finished programming events, set the MODE switch to RUN.

Setting the Wake Up Alarm

When you set the Mini Timer to control a module with Unit Code 1 or 5, you can set a buzzer to go off as well. To do this, create an event for Unit Code 1 or 5 by following the instructions above. Then set the WAKE UP switch to IN. You can deactivate the wake-up buzzer by setting the WAKE UP switch to OUT.

You should consider the Wake Up Alarm feature when you decide which modules you want to assign Unit Code 1 or 5 and which SELECTOR setting you want to use for the Mini Timer. For example, if you plan to use the Wake Up Alarm, you should assign Unit Code 1 or 5 to devices that you want to turn on when you wake up, such as a coffee pot. You should assign these devices Unit Code 1 or Unit Code 5, depending on how you have set the SELECTOR switch.

Note that when the buzzer sounds, pressing any button on the Mini Timer activates the "snooze" feature, which turns the buzzer off for 10 minutes. After this 10-minute period, the buzzer sounds again. It goes off automatically after it has sounded for 10 minutes. If you want to turn the buzzer off permanently, just set the WAKE UP switch to OUT.

You can temporarily suspend timed events for modules with Unit Code 1 or 5 by setting the MODE switch to CLOCK SET and the WAKE UP switch to OUT. This is useful when you want to turn off timed events for a short period of time. For example, if you normally have the light in your bedroom turn on at 6:00 AM, you can suspend this event after a late night simply by setting the MODE switch to CLOCK SET and the WAKE UP switch to OUT. Note that when you do this, you can still use the rocker buttons on the front of the Mini Timer to control devices manually.

At this point, you know all you need to know to set up timed events with the Mini Timer. The sections that follow give details on advanced features of the Mini Timer. Before reading these sections, you might want to try out what you've already learned.

One-Time Timed Events

Sometimes you might want to set a timed event to happen only once. For example, you might set the porch light to turn on at midnight on a particular day because you know you're going to be home late. You do this by using the Once mode.

To specify that an event happen only once, press ONCE within 4 seconds of setting a timed event.

The timed event will occur at the specified time within the next 24-hour period, and then be cleared from the Mini Timer's memory.

Security Mode

One of the benefits of being able to set timed events is that you can give your house that "lived in" look, even when you're not at home. By using the Security mode, you can enhance the lived-in look by having timed events occur at slightly different times every day. For example, if you set the porch light to go on at 6:00 PM every day, and then set Security mode, the light will come on at 6:00 PM the first day, and then at a random time between 6:00 PM and 7:00 PM every day after that.

To activate Security mode, press SECURITY within 4 seconds of setting a timed event.

Sleep Mode

Occasionally, you might want to turn a module on or off after a specific amount of time. For example, you might want to listen to the radio while falling asleep, and have the radio turn off automatically after 15 minutes. You can do this by using the Sleep mode. Here's how to use Sleep mode to turn a module off:

Press ON for the module you want to control. Then, within 4 seconds, press SLEEP once for each 15 minutes you want the module to remain on before turning off.

For example, assume that your radio is connected to an Appliance Module set to Unit Code 1. If you want the radio to play for 30 minutes and then turn off, press ON for the rocker button marked "1•5," and then press SLEEP twice within 4 seconds (2 x 15 minutes = 30 minutes).

You can also use the Sleep mode to turn a device on. Here's how:

Press OFF for the module you want to control. Then, within 4 seconds, press SLEEP once for each 15 minutes you want the module to remain off before turning on.

For example, assume once more that your radio is connected to an Appliance Module set to Unit Code 1. If you want to take a nap for 45 minutes and then have the radio turn on, simply press OFF for the rocker button marked "1•5" and then press SLEEP three times (3 x 15 minutes = 45 minutes) within 4 seconds.

Sleep mode is the last of the "advanced" Mini Timer features. The rest of this section covers reviewing, modifying, and deleting timed events.

Reviewing Timed Events

There are two reasons you'll need to know how to review timed events: to confirm the events that you've already set, and to clear previously defined events.

First, you'll learn how to review the events that you've already set; but to do that, you'll need to set the mode.

Set the MODE switch to PROG REVIEW.

PROG REVIEW, as you may have guessed, stands for "Program Review." Now you can check events for each of your devices.

Press either ON (to review On events) or OFF (to review Off events) for the rocker button of the device whose timed events you want to confirm.

For example, if you want to check the time when the module or modules with Unit Code 1 will go on, press ON for the rocker button marked "1•5." After you press the button, the time of the event is displayed. If there is no event set for that Unit Code, "0:00" is displayed. Note that if two events are set (for example, if the module is set to go on at 6:00 AM and 6:00 PM), you'll need to press the Unit Code button twice to see both times.

You can review events for more than one Unit Code simply by pressing more rocker buttons. When you're finished, you need to change the mode back to RUN, or the Mini Timer won't execute your timed events.

Once you have reviewed the settings you are interested in, set the MODE switch to RUN.

Clearing Events

Sometimes you'll want to clear events from the Mini Timer's memory to make room for new timed events, or simply because you don't want the old events to be active any longer. To clear events, you first have to change the mode.

Set the MODE switch to PROG REVIEW.

Next you select the event you want to clear.

Select the event you want to clear by pressing ON or OFF for the appropriate rocker button until the event is displayed.

Note that if two events are set for the Unit Code whose event you want to clear, you need to ensure that the correct one is displayed before continuing; otherwise, you'll clear the wrong event.

Once the event is displayed, it's easy to clear.

Press CLEAR to remove the event from the Mini Timer's memory.

You can repeat this procedure for as many events as you want to clear. When you are finished, you'll need to reset the mode.

When you have finished clearing events, set the MODE switch to RUN.

Remember that you can create only two On events and two Off events for each Unit Code. If you try to create a timed event when there are already two events programmed for that Unit Code, the display will show "18:88." Before creating this new event, you must use the procedure outlined above to remove one of the existing events.

Installing the Backup Battery

If you do not install a backup battery in your Mini Timer, and the power goes off for some reason (say, during a thunderstorm), you will lose the clock setting and any timed events that you've set.

Here's how to install the backup battery:

Open the battery compartment cover.

The battery compartment cover is located on the back of the Mini Timer.

Connect a 9-volt battery to the battery contact inside the compartment. Then replace the battery compartment cover.

Once you've installed the battery, look at the front of the Mini Timer. To the right of the display, there is an LED indicator labeled "Battery Sentinel." When the LED comes on, your battery is weak and should be replaced.

Comparing the Mini Timer with Other Controllers

The key feature of the Mini Timer is that it supports timed events. If you need to turn devices on and off automatically at specific times during the day or night, then the Mini Timer might be for you. The key question to ask yourself is "Are there devices that I want to turn off or on at the same time *every* day?"

Here are some situations that might require the functionality of the Mini Timer:

- Turning on an outside porch light at dusk every evening
- Turning on a coffee maker first thing every morning
- Turning "grow lights" on and off at the same time every day
- Controlling sprinkler systems, outdoor lights, or other low-voltage electrical devices
- Turning a "bug zapper" on every night and off every morning

If you decide you need a controller that supports timed events, you'll have to choose between a Home Automation Interface and a Mini Timer. (Well, you don't *have* to choose: I own three Home Automation Interfaces and one Mini Timer!) The biggest factor in your decision should be whether you have a personal computer. If you do, you should almost certainly purchase the Home Automation Interface.

The Home Automation Interface can control modules with any address—that is, any combination of Housecode and Unit Code. The interface can store up to 128 timed events. In comparison, the Mini Timer can control only modules with the same Housecode it uses, and it can control only one group of four modules with that Housecode: either those with Unit Codes 1 through 4, or those with Unit Codes 5 through 8. In addition, it can store only two On and two Off events for each module.

Note that in the previous paragraph, we used the term *modules* to mean modules with unique addresses. Naturally, you can control more than four modules with the Mini Timer if some of them share the same Unit Code. For example, you could control any number of lights if each of them were connected to a module set to Unit Code 1.

Home Automation Interface

Overview

The Home Automation Interface is used with a personal computer to control up to 256 X-10 modules (or more, if some modules share the same address). It comes with software and a cable for connecting it to your computer. Detailed information on using the interface with a Macintosh computer is given in Chapter 6, "Apple Macintosh," on page 93. Information on using the interface with an IBM PC or compatible is given in Chapter 7, "IBM PC," on page 133. Versions are also available for the Apple II and the Commodore 64 and Commodore 128 computers.

Here is a diagram of the interface, with a description of the major components.

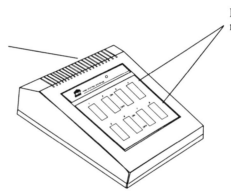

Attach the interface to your computer and use the software that comes with it to set timed events or to control devices immediately. (A cable is also included with the interface.)

Press the rocker buttons to turn modules on and off manually.

Using the Interface for Manual Control

You can use the interface to control up to eight modules manually simply by pressing a rocker button. You control modules manually with the interface the same way that you control them with the Mini Controller. If you are not sure how the rocker buttons work, refer to the instructions for the Mini Controller on page 14.

Using the Interface for Timed Events

The real value of the interface comes when you use it with your computer. The software that is included with the interface allows you to set timed events for up to 256 modules.

Unlike the Mini Timer, the interface lets you set timed events for modules that have *different* Housecodes. This gives you maximum flexibility.

For example, you can use the interface to create timed events for lights attached to Lamp Modules that use Housecode A, as well as for sprinklers attached to Universal Modules that use Housecode B. What's more, with a Macintosh computer, you can use a graphics package to create a floor plan of your house, and then use the floor plan in conjunction with the interface software. This makes it easier to identify all the modules in your system, and to review the timed events you've set for each module.

Comparing the Interface with Other Controllers

Like the Mini Timer, the big advantage of the Home Automation Interface is the ability to create timed events.

If you don't have a personal computer, the biggest drawback of the interface is certainly the requirement that you gain access to a PC to use it. If you already have a computer, though, the only drawback of the interface is that it's more expensive than the Mini Timer (but probably not too much more—shop around!).

The interface is definitely the most flexible controller described in this book: Using the software that comes with it, you can set up to 128 timed events, and each of these timed events can correspond to a module with a unique address. That's a lot. In addition, the interface is the only controller that allows you to control modules that have different Housecodes.

The interface makes an ideal foundation for almost any home-automation system. You could use the interface to set up timed events for *all* the modules in the system, and then use other controllers to add functionality. For example, you could use a Telephone Transponder to control devices when you're away from home, and you could use a Remote Control to turn the TV off without leaving your favorite chair.

Telephone Transponder

Overview

The Telephone Transponder lets you control up to eight modules manually, using the rocker buttons on the front of the transponder. In addition, the Telephone Transponder can be used to control up to ten devices remotely from *any* Touchtone telephone in the United States.

A diagram of the Telephone Transponder is shown below, with a description of the major components.

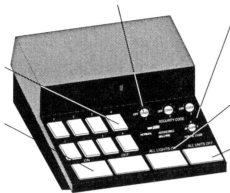

Set one or more of these dials to establish a security code. Enter the security code first when calling the Telephone Transponder from a remote location.

Set the Housecode here. Make sure it matches the Housecode on the modules you want to control.

Click one or more of the eight Unit Code buttons to identify the Unit Code(s) of the modules you want to control.

After pressing one or more Unit Code buttons, press ON or OFF to control the selected modules.

ALL LIGHTS ON turns on all lights with the same Housecode as the Telephone Transponder, as long as the lights are connected to Lamp Modules or Wall Switch Modules.

ALL UNITS OFF turns off all devices attached to modules with the same Housecode as the Telephone Transponder, including devices attached to Appliance Modules.

Using the Telephone Transponder for Manual Control

To control devices in your home manually, use the Telephone Transponder exactly as you would use the Maxi Controller. In case you've forgotten, here are the basic steps:

- Plug the Telephone Transponder into any electrical outlet.
- Set the Housecode on the Telephone Transponder so it matches the Housecode on the module or modules you want to control.
- Select the module or modules you want to control by pressing the appropriate Unit Code buttons.
- Choose the function command that you want to send to the modules by pressing ON or OFF.

- Press ALL LIGHTS ON to turn on all lights attached to modules with the same Housecode as the Telephone Transponder. (ALL LIGHTS ON turns on only lights attached to Lamp Modules or Wall Switch Modules.)
- Press ALL UNITS OFF to turn off all devices attached to modules with the same Housecode as the Telephone Transponder, including devices attached to Appliance Modules.

That's it! If things are still a little unclear, please see "Using the Maxi Controller" on page 10.

Using the Telephone Transponder for Remote Control

Now you know how to use the Telephone Transponder to control devices manually. The really nifty feature of the Telephone Transponder, though, is its ability to control devices remotely, from any telephone in the United States. The following sections tell you how.

Setting the Security Code

You may want to set the security code before attaching the Telephone Transponder to your phone system. This helps to prevent ne'er-do-wells from calling and controlling devices in your house. It also prevents the unsuspecting from accidently turning devices on or off in your home when they call and end up talking to the Telephone Transponder instead of to you.

To set the security code, use a small screwdriver to turn the three dials marked "SECURITY CODE" on the face of the Telephone Transponder.

You can set a one-, two-, or three-digit security code. If you decide to set a one-digit code, be sure to use the Security Code dial labeled "1." If you decide on a two-digit code, use dials 1 and 2.

The diagram below shows how the dials would be set for no security code, for a single-digit code, and for a three-digit code.

No security code set. When you call, enter commands as soon as the Telephone Transponder answers the phone with three beeps.

Here, the security code is 5. When the Telephone Transponder answers the phone with three beeps, press 5, then enter commands.

Here, the security code is 8-6-5. When the Telephone Transponder answers the phone with three beeps, press 8-6-5, then enter commands.

Setting Up the Telephone Transponder

You're ready to set up the Telephone Transponder. By now the first step should be obvious.

Plug the Telephone Transponder into any electrical outlet.

The only other installation required is connecting the Telephone Transponder to a phone line so it can accept incoming calls.

Connect the Telephone Transponder to a standard phone line by plugging the modular phone jack connector attached to the Telephone Transponder into any modular telephone jack.

This is **techno-speak** for "plug the Telephone Transponder into a telephone jack." If you don't have a spare telephone jack, you can either have one installed or buy a "coupling jack" to connect both the phone and the Telephone Transponder to the same telephone jack. You can find a coupling jack at most stores that sell telephones.

Setting the ANSWERING MACHINE Switch

If you have an answering machine, you can still use the Telephone Transponder, but you need to tell the transponder to expect an answering machine to answer the phone.

Set the switch in the center of the Telephone Transponder to ANSWERING MACHINE if you have one. Set the switch to NORMAL if you don't have an answering machine.

Operating the Telephone Transponder is a little different if you have an answering machine; the details are covered in the next section.

Remote Operation

To use the Telephone Transponder, the first thing you do is leave the house and phone home.

Call the phone number of the phone line to which you've attached the Telephone Transponder. Then wait for the Telephone Transponder to answer the phone.

When the Telephone Transponder answers, you will hear three short beeps.

If you have an answering machine, and you've set the ANSWERING MACHINE switch accordingly, the Telephone Transponder will sound three short beeps 30 seconds after the first ring, which will probably be *after* your answering machine message finishes. Be patient!

If you set a security code, enter it now. Otherwise, skip to the next instruction.

The Telephone Transponder will not do anything after you enter the security code.

Press the number on the phone keypad that corresponds to the Unit Code of the module or modules you want to control. Then press ✻ to turn the devices on, or press # to turn the devices off.

Note that you can control only modules with Unit Codes from 1 to 10. To control a module with Unit Code 10, press 0. After you enter a command, you should hear the characteristic three short beeps of the Telephone Transponder. After the three beeps, you can either enter another command or hang up the phone.

32   Controllers

Flashing the Lights

There's one more feature of the Telephone Transponder that you should know about. Any lights connected to modules set to the same Unit Code as the third Security Code dial will flash on and off when the phone rings. This is a very useful feature for those who are hearing impaired, because they can have the lights flash on and off when the phone rings. Even if you don't have a hearing impairment, this feature can be useful if you don't want to miss phone calls while you're outside at night, or when the stereo is playing loudly.

That's it! You know how to use all the features of the Telephone Transponder.

Comparing the Telephone Transponder with Other Controllers

You'll need a Telephone Transponder if you want to control devices when you're away from home. Here are some examples:

- To turn on a heater or air conditioner so the house is comfortable when you get home
- To ensure that appliances, such as a heater or an iron, are off when you're away from home
- To turn lights on for added security when you are away from home
- To turn sprinklers on and off when you're away from home

Because it lets you control devices remotely, the Telephone Transponder is ideal for people who have vacation homes. With the Telephone Transponder, you can adjust the heating or air conditioning in the vacation home before you arrive, and you can also ensure that you've turned off all the appliances and lights when you leave.

If you don't need the ability to control devices while you're away from home, you should probably purchase a different controller with features more suited to your needs.

Remote Control and Wireless Transceiver

Overview

The Remote Control is used with the Wireless Transceiver to control up to 16 devices from various locations around your home.

Here's a diagram of the Remote Control and the Wireless Transceiver, with a description of the major components.

Remote Control

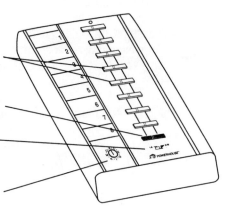

Press the left side of a rocker button to turn the device on. Press the right side to turn the device off.

After turning a lamp on, press this button to dim or brighten the lamp.

Set the SELECTOR switch to control modules 1 through 8 or 9 through 16.

Set the Housecode for the Remote Control here. Make sure it matches the Housecode on the Wireless Transceiver and on the modules you want to control.

Wireless Transceiver

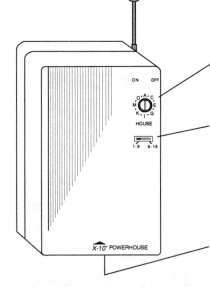

Set the Housecode for the Wireless Transceiver here. Make sure that it matches the Housecode on the Remote Control and on the modules you want to control.

Then set the Unit Code for the Wireless Transceiver. It will respond to commands sent to this Unit Code, and will control an attached device accordingly.

Plug a lamp or appliance into the two-prong outlet, and control it by using the Wireless Transceiver as an Appliance Module (see page 47).

You can use the Remote Control just as you would a remote control for a TV. There's one difference, though. To use a TV remote, you must be within sight of the TV, and you must point the remote directly at the unit. The X-10 Remote Control uses technology that allows you to control devices from greater distances, without having to point the remote directly at the Wireless Transceiver. That is, you can carry the Remote Control with you around the house and turn devices on and off simply by pressing a button. You could even use the Remote Control to open your garage door and turn on the house lights from your car as you enter the driveway!

Using the Remote Control

Install four AAA batteries in the Remote Control. Then set the Housecode on the Remote Control to the same Housecode you've set on the Wireless Transceiver.

Note that the Housecode you choose must also be the Housecode of any modules you want to control with the Remote Control.

Plug the Wireless Transceiver into any standard electrical outlet, and set the SELECTOR switch to control modules 1 through 8 or 9 through 16.

You can control either modules 1 through 8 or modules 9 through 16, but you cannot control modules from both groups.

If you want, you can also plug an appliance into the Wireless Transceiver. If you do, the transceiver will behave like an appliance module with Unit Code 1 or 9, depending on how you have set the SELECTOR switch.

Set the SELECTOR switch at the bottom of the Remote Control.

Move the switch to the left to control modules with Unit Codes 1 through 8; move the switch to the right to control modules with Unit Codes 9 through 16. Note that the SELECTOR setting on the Remote Control *must* match the SELECTOR setting on the Wireless Transceiver.

Use the rocker buttons on the Remote Control to turn devices on and off.

The rocker buttons are labeled 1 through 8. Press the button that corresponds to the Unit Code of the module you want to control. Press the left side of the rocker button to turn the device on. Press the right side to turn the device off.

If you are controlling a light attached to a Lamp Module or a Wall Switch Module, you can also use the dimmer button to dim or brighten the lamp to the intensity you want.

The dimmer button is near the bottom of the Remote Control. If you look closely, you'll see two arrows embossed on it. Press the left side of the button to brighten the light; press the right side to dim the light.

Comparing the Remote Control with Other Controllers

The Remote Control is useful for someone who has trouble getting around. It's also a great solution for turning lights on from outside the house, even from the car as you're pulling into the driveway.

You could also use the Remote Control from your backyard to turn sprinkler systems on and off. This is very helpful if you have drip systems and you need to check for faulty components by turning the sprinklers on and off frequently.

For information on using the X-10 Universal Module to control sprinkler systems and other low-voltage electrical devices, see the section "Universal Module" on page 57.

The Remote Control could be used as the only controller in your system, especially if you don't need to create timed events. It has all the functionality of a Maxi Controller, and more: It's transportable. But remember, there's no reason to limit yourself—you can always use *both* a Maxi Controller and a Remote Control in your home-automation system.

3 Controlling Lights

Overview

In this chapter, we'll show you how you can use X-10 modules to control the lights in your home. With an X-10 module connected to a lamp or replacing a light switch, you can turn the lights on or off, and dim them to whatever brightness you want, from anywhere in the house. For example, you can turn lights on in the garage without leaving your car, or turn off all the lights in your home from a single button at your bedside, or increase security when you're away by turning lights on and off at various times during the day. More suggestions are given in the section "Applications" on page 45.

Although many variations exist, there are three main X-10 modules that you'll use to control lights in your home or business: the Lamp Module, the Two-Way Wall Switch Module, and the Three-Way Wall Switch Module. Each module is described in this chapter.

By the end of the chapter, you'll be able to choose and install the best type of module for the lights in your house.

Lamp Module

The Lamp Module is used to control the incandescent, free-standing lamps in your house. After plugging a lamp into the module and the module into the wall socket, you can use any X-10 controller to turn the lamp on and off, and to dim the lamp to any brightness you want.

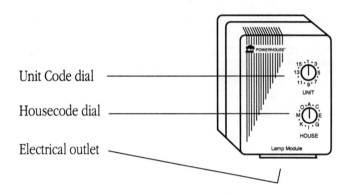

Unit Code dial

Housecode dial

Electrical outlet

There are a few restrictions you should observe when using the Lamp Module.

The Lamp Module should be used only with **incandescent** lamps. Most of the lamps in your home are probably incandescent. You cannot use the Lamp Module to control **fluorescent** lamps. You can generally identify the lamps in your house by the type of bulbs they use. Incandescent lamps use standard, well, bulb-shaped bulbs; fluorescent lights generally use either long, thin tubes or smaller, circular tubes. Another way to tell is that incandescent bulbs generate heat. They'll burn you if you touch them after they've been on awhile. Fluorescent bulbs in good working condition generate very little heat.

The Lamp Module works with lights that use up to 300 **watts** (w) of electricity. Check the bulb in your lamp—it should display its power consumption. If a lamp uses two or more bulbs, you'll need to add the wattage of the bulbs to determine whether the Lamp Module can be used to control the lamp. For example, you could use the module to control a lamp with two 150w bulbs (300 watts total), but not to control a lamp with two 200w bulbs (400 watts total). This restriction also applies if two lamps are plugged into a single module. Calculate the total wattage of all bulbs controlled by a single module to make sure you don't exceed this limit.

The Lamp Module has no minimum wattage constraint and is easy to install and use.

Set the Housecode dial on the module to the same letter as the Housecode on the controller you'll use with it.

This step ensures that the Lamp Module and the controller will communicate correctly. As your system gets more sophisticated, you'll use more than one Housecode. For now, just use "A."

Set the Unit Code to an unused number from 1 to 16. Since this is your first module, use "1."

By setting a unique Unit Code for each module, you will be able to control the lamp that is attached to the module. If you set two Lamp Modules to the same Unit Code, they will be controlled at the same time with the same commands.

Plug your lamp into the bottom of the Lamp Module.

Plug the Lamp Module into an unswitched wall outlet.

An unswitched wall outlet is simply a wall outlet that is not controlled by a wall switch of any type. The Lamp Module can work with a switched outlet, but it won't work if you turn the switch off, and we don't recommend it.

Make sure any switch on the lamp itself remains in the ON position.

An X-10 Lamp Module can't control a lamp that has been turned off at its switch.

That's it! Try it out:

Make sure your controller is plugged into an unswitched outlet. Push "1" and then ON.

Some controllers have a single rocker switch to select a Unit Code and turn the module on or off. Refer to Chapter 2 if you've forgotten how they work.

The lamp should come on.

Now press "1." Then press DIM and hold it.

The lamp should gradually become dimmer as you hold down the DIM button. Try BRIGHT and OFF. Lamp Modules respond to ALL LIGHTS ON and ALL UNITS OFF commands as well.

Another feature of X-10 modules is their "local control" capability. If you are standing next to your lamp and don't want to walk to the controller, you can simply turn the lamp's switch off and then on again, and the lamp will turn on. Make sure to leave the switch on so you can continue to control it from a controller.

If things aren't working, make sure you are using the same Housecode on the module as on your controller, and that you've set the Unit Code to "1." Make sure the outlets you're using are unswitched. If that doesn't help, see Appendix B, "Troubleshooting," on page 163.

Two-Way Wall Switch Module

The Two-Way Wall Switch Module performs the same functions as the Lamp Module, but replaces a built-in wall light switch, allowing you to remotely turn the lights on and off or dim them. The Wall Switch Module is rated at 500w maximum, 60w minimum, and can be used only with incandescent lights. This module can also function as a regular light switch to turn lights on and off (you must use a controller to dim the lights).

The slide switch under the main push button turns off the connected lights and also turns off the X-10 module so the light can't be turned on or off remotely, a safety feature required by the Underwriters Laboratories. You also use the slide switch to keep a light off or to turn off the power to the light before you change a light bulb.

You'll use the Two-Way Wall Switch Module when only one existing wall switch controls a light. If two switches can control a light (one by each door, perhaps), you'll need to use the Three-Way Wall Switch. If only one switch controls a light or lights, it is connected to the circuit by two wires. If a switch is connected by three wires, it is usually a three-way wall switch.

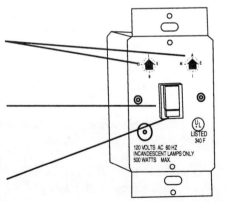

Set the Housecode and Unit Code for the Wall Switch Module.

You can use the Wall Switch Module like a standard light switch: Press once to turn lights on; press again to turn them off.

A slide switch is located under the main push-button switch. Slide this to the left to turn the switch off so it can't be turned on by a controller.

Installing a Two-Way Wall Switch Module

 WARNING! Installing a Wall Switch Module requires knowledge and experience in electrical wiring. If you are not familiar with electrical wiring, please consult a qualified electrician. Household electricity can be lethal.

Use a small screwdriver to set the Housecode to the same letter you are using on your controller. To begin with, use "A."

The Wall Switch Module is now ready to receive commands from your controller.

Set the Unit Code to an unused number from 1 to 16.

It's a good idea to make a list of the Housecode and Unit Code numbers for all your modules, so you can keep track of them as your system grows.

Turn off the power at the circuit breaker box or fuse box.

Remove the existing wall plate and the switch from the electrical box.

If more than two wires are connected to the switch, there's a good possibility that you have a three-way wall switch, that is, a light that is controlled by more than one wall switch. If there are more than two wires, leave the wires connected and put the switch back together. Then proceed to the next section.

Remove the two wires from the switch.

Connect the two wires on the Wall Switch Module (black and blue) using the wire nuts provided.

Make sure all of the exposed metal wire is covered by the wire nut.

Reinstall the Wall Switch Module in the electrical box.

Make note of the Housecode and Unit Code.

Replace the cover plate.

Turn on the power at the main circuit breaker.

That's it!

To test your Wall Switch Module:

Slide the switch under the push button all the way to the right.

When you move this slide switch to the left, it turns off the light and turns off the Wall Switch's ability to receive commands from a controller. Use this feature whenever you're replacing the bulb. Sometimes this slide switch is inadvertently turned off, so this is the first place to look when troubleshooting your Wall Switch Module.

Press the Unit Code for the Wall Switch Module and the ON button on your controller.

Let there be light!

Push the push button on the Wall Switch Module to check local operation.

One push of the button will turn the light off, and the next will turn it on. Dimming is available only at the controller, not at the switch itself.

If you are having problems, please see Appendix B, "Troubleshooting," on page 163.

Three-Way Wall Switch Module

The Three-Way Wall Switch Module can be used wherever lights are controlled by two or more switches. It includes two units that replace all the switches that currently control a single set of lights. The Three-Way Switch Module is also rated at 500w maximum, 60w minimum. The maximum is calculated the same way as for the Two-Way Wall Switch.

The Three-Way Wall Switch Module comes with two light switches, which replace two switches you are currently using to control a single light.

One switch contains an X-10 module that receives commands. This is called the Master Switch. The other acts as a normal three-way switch; it's called the Companion Switch. They must be used together.

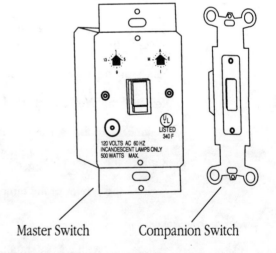

Master Switch   Companion Switch

42 Controlling Lights

If you have a light that is controlled by more than two existing wall switches, you can purchase additional Companion Switches. Only one Master Switch is needed no matter how many wall switches are included on the circuit.

Installing a Three-Way Wall Switch

Installing a Three-Way Wall Switch Module requires knowledge and experience in electrical wiring. Instructions are given here for a couple of common wire color codes. Your house may or may not follow these conventions. If you are not familiar with electrical wiring, please consult a qualified electrician. Household electricity can be lethal.

Use a small screwdriver to set the Housecode to the same letter you are using on your controller. To begin with, use "A."

The Three-Way Wall Switch Module is now ready to receive commands from your controller.

Set the Unit Code to an unused number from 1 to 16.

It's a good idea to make a list of the Housecode and Unit Code numbers for all your modules, so you can keep track of them as your system grows.

Turn off the power at the breaker box.

WARNING! Household electrical current can be deadly. Hire an electrician if you're not completely familiar with the installation of electrical switches.

Select one of the existing switches. Either switch is fine, but if one of the switches is in an electrical box by itself, choose that one to install the X-10 Master Switch. Since the Master Switch is larger than the Companion Switch, it requires more room in the electrical box.

Remove the existing wall plate and wall switch from the electrical box.

Write down the color of each wire and where it's connected.

If something doesn't work, it's always nice to put things back the way they were.

On the existing switch, remove the wire that is connected to terminal that is a different color from the other two terminals.

The terminals are the screws that the wires attach to.

Connect this wire to the blue wire on the Master Switch, using a wire nut.

No bare metal should be exposed.

Remove the other two wires, called "travelers," from the existing switch. If one of them is red, connect it to the red wire on the Master Switch.

Even if there are no red wires connected to the existing switch, write down the color of the wire you connect to the X-10 Master Switch. It will be important when you connect the Companion Switch.

Install the Master Switch in the electrical box, but leave off the cover plate.

Now you'll install the smaller, Companion Switch.

Remove the existing wall plate and switch from the electrical box.

Write down the color of each wire and where it's connected.

On the existing switch, remove the wire that is connected to a terminal that is a different color from the other two terminals.

Connect this wire to one of the two blue wires on the Companion Switch, using a wire nut.

No bare metal should be exposed.

Connect the same-color wire to the red wire on the Companion Switch as you connected to the red wire on the Master Switch.

Connect the remaining wire on the existing switch to the second blue wire on the Companion Switch.

If there is a discrepancy in the wiring, a different standard may have been used—or worse, the wiring in your house may be improperly installed. If your wiring is different from this description reinstall the old switches and call an electrician.

If everything checks out:

Install the Companion Switch in the electrical box.

Install both cover plates.

Move the small slide switch under the push button on the Master Switch to the center, or on, position.

Turn on the power at the main circuit breaker or fuse box.

That's it!

To test your Three-Way Wall Switch Module:

When the slide switch is moved to the left, it turns off the light and turns off the Three-Way Wall Switch's ability to receive commands from a controller. Use this feature to make sure someone doesn't turn on the light remotely while you're replacing a light bulb. Sometimes this slide switch is inadvertently turned off, so this is the first place to look when troubleshooting your Three-Way Wall Switch.

Press the Unit Code for the Wall Switch and the ON button on your controller.

Let there be light!

Push the push button on the Three-Way Wall Switch Module and on the Companion Switch to check local operation.

One push of the button will turn the light off, and the next will turn it on again. Dimming is available only at the controller, not at the switch itself.

If you have problems, please see Appendix B, "Troubleshooting," on page 163.

Applications

Now that you've got everything installed, you may already be thinking of ways to use your X-10 light-control system. To help you on your way, here are a few suggestions and strategies.

Before you start to hook up your Lamp Modules, think about the way you use lights in your house. Which ones are used most frequently? Which light switches are the farthest from the light source? When do you find yourself walking through a dark room to get to a light switch? Which room in your house is the most uncomfortable to walk through in the dark? Are there lights you'd like to dim? The answers to questions like these will help you to decide which lights to add modules to first.

Security. You can increase the security of your house with your X-10 home-control system. Use an X-10 Mini Timer or Home Automation Interface to turn the lights on and off at intervals while you're away from home. You can also use any X-10 controller with an "all lights on" button to turn on all the lights in your house from your bed if you suspect an intruder. With special controllers described in Chapter 5, you can have all the lights in the house flash if entry or movement is detected. This can thwart a robbery before it ever happens—the best kind of protection.

Safety. Use an X-10 controller to switch lights off remotely. For example, turn basement or garage lights on or off without walking downstairs or outside in the dark. With a Wireless Remote Control System, you can turn on selected lights in your house as you drive up in your car. Use a Mini Timer to turn your front-porch light or pool light off and on automatically every night—even when you're out of town—to avoid dangerous situations on your property, and possible liability for negligence.

Convenience. You can also use Lamp Modules and a Mini Timer or Home Automation Interface to turn plant grow lights or aquarium lights on and off every day.

Sometimes, it might make sense to turn a particular light *off* at a certain time every day. For example, you might have an outdoor light or a game-room light that no one ever seems to remember to turn off. You can solve the problem by setting an X-10 Mini Timer to turn the light off every night at a certain time, say midnight, after everyone is normally asleep. And if the light's already off when the Mini Timer tries to turn it off, that's not a problem—the module will simply ignore the command.

4 Controlling Appliances

Overview

This chapter will give you an overview of the X-10 modules available to control appliances and help you decide which ones might be best for you. Appliances are generally described as all "nonlight" electrical devices. Typical applications include coffee pots, air conditioners, heaters, attic fans, and pool equipment. Appliance Modules are available for 110-volt, two- and three-pin plugs, as well as for 220-volt appliances.

Generally, 220-volt systems are used for appliances with high energy demands, such as air conditioners, shop equipment, pool filters, and the like. Appliances that use 220 volts can be identified by the type of plug they use. In the United States, 220-volt systems usually have plugs with three prongs that are angled differently than on those standard 110-volt systems. As always, if you have any questions, consult an electrician.

A device called the Thermostat Set-back Controller works in conjunction with an Appliance Module to control virtually any central heating and air-conditioning system.

There is also a special type of module, called a Universal Module, for appliances that operate on **low voltage** (up to 30 volts), such as sprinkler systems, low-voltage electric motors, and low-voltage outdoor lights.

All of these modules are covered in this chapter. When you've finished it, you'll be able to choose the module you need and install it in your house.

Modules

Appliance Modules are different from Lamp Modules and shouldn't be interchanged. Appliance Modules don't respond to the DIM command. This is extremely important to know, because most appliances don't respond well to the low voltages that result when a DIM command is received by a Lamp Module. When a DIM command is sent, the module lowers the amount of electricity delivered to the appliance. Electric motors can be severely damaged if they are run on less than the appropriate voltage. This problem is not unique to X-10 systems—you should never operate an appliance on any type of dimmer switch.

You must also be sure that a remotely controlled heating appliance does not come in contact with flammable material. Fire could result. It is also possible to turn on an appliance remotely without realizing it, and to leave it on for an extended period of time. This too is a dangerous situation. Don't let these potential problems discourage you from using Appliance Modules, though. With a little planning and common sense, you can ensure that your system will be safe.

For example, if you turn on your coffee maker at 7:00 AM with a Mini Timer or Home Automation Interface, be sure to send an OFF command at 9:00 AM. If you remember to turn off the coffee maker manually, the command will have no effect. If you forget to turn it off, the X-10 system will do it for you. You could also use the Telephone Transponder to call home from wherever you are and put your mind at ease by sending an OFF command to the module. A well-planned X-10 system can help prevent accidents in your home.

Two- and Three-Pin Appliance Modules

The X-10 Appliance Module is rated at 15A resistive load. This is the equivalent of about 1800 watts for items such as coffee pots, slow cookers, electric blankets, and the like. Check the manufacturer's tag on your appliance to ensure that it's within the guidelines. A typical appliance, such as a coffee maker, might have something like "120V≈/900W" stamped on the bottom or listed in its manual. This indicates that the coffee maker runs on standard household current of 120 volts and draws 900 watts of power, about half the 1800-watt total capacity of an Appliance Module.

An Appliance Module can also be used to control lights up to a 500-watt total capacity. The 1800-watt rating cannot be translated to more capacity for large or multiple lights. Due to the nature of how incandescent lamps "start up" an Appliance Module can be used to control lamps up to a 500-watt total capacity. Note that when you use an Appliance Module to control an incandescent lamp, the module will not respond to DIM or ALL LIGHTS ON commands.

Appliance Modules can also be used to control fluorescent lamps, but again, they do not respond to the DIM and ALL LIGHTS ON commands.

The X-10 Appliance Module is rated at 1/3 horsepower for electrical motors. This is sufficient for most fans, sump pumps, air conditioners, and similar devices, but be sure to check the listing for each appliance you plan to control.

It is very important that you check each appliance against the rated capacity of the module before installing it. Fire could result if the modules are severely overloaded.

There are Appliance Modules with two-pin and three-pin grounded connections. Both types work with standard 110-volt appliances and have a polarized connection (that means you can plug in the appliance in only one way). You can use a two-pin appliance in a Three-Pin Appliance Module, but never cut off a ground pin on a three-pin appliance in order to use it with a Two-Pin Appliance Module.

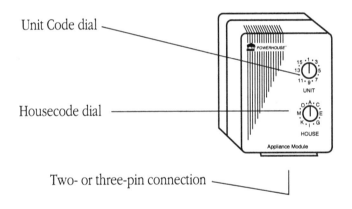

Unit Code dial

Housecode dial

Two- or three-pin connection

Appliance Modules are very easy to set up:

Set the Housecode to the same setting as the Housecode on the controller you plan to use. For now, use "A."

Set the Unit Code to a unique number (one that isn't being used by any other module).

Plug the appliance into the module (the outlet is on the bottom of the module).

The connection is "polarized," meaning that some appliance plugs will fit in only one way. If the plug doesn't go in easily, turn it over.

Plug the Appliance Module into a nonswitched outlet (one that isn't controlled by a wall switch).

Bingo!

Test it out with your controller. Make sure the appliance's local switch is on.

Press ON for the Unit Code that you assigned.

On comes the appliance.

Another feature of X-10 modules is their "local control" capability. If you are standing next to your appliance and don't want to walk to the controller, you can simply turn the appliance's switch off and then on again, and the appliance will turn on. Make sure to leave the switch on when you're done so you can continue to turn it on and off from your controller.

Wall Receptacle Module

The Appliance Module described above lets you control most appliances, but it isn't always esthetically pleasing to have the module protruding from the wall. The Wall Receptacle Module incorporates the same electronics, but they are housed in a replacement receptacle that fits in a standard electrical box. Except for the Housecode and Unit Code dials on the front, the module looks exactly like a regular outlet. This module provides one outlet (on the top) that is controlled via X-10 and one outlet (on the bottom) that acts like a standard electrical outlet. The bottom outlet is not affected by any X-10 commands.

The Wall Receptacle Module is rated at a full 15A (1800-watt) load, but unlike the Appliance Module, it is not restricted to 500 watts for lamps. It can also handle a 2-horsepower motor.

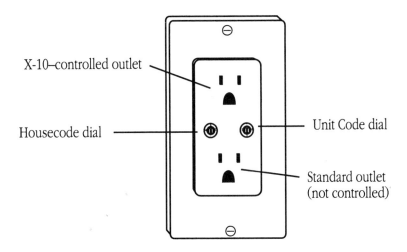

 Installing the X-10 Wall Receptacle Module requires knowledge and experience in electrical wiring. Household electricity can be lethal. If you are not absolutely sure you know what you are doing, consult an electrician.

Turn off the power at the circuit breaker or fuse box.

Remove the existing wall plate and outlet from the electrical box.

Write down the colors of the wires and where they are connected.

If something doesn't work, it's always nice to know how to put it back together.

Connect the white and black wires on the Wall Receptacle Module, black to black and white to white, using the wire nuts.

Make sure no bare metal is exposed on the wires.

Connect the green wire to the green or bare ground wire or to the metal box itself.

Install the Wall Receptacle Module in the electrical box and replace the screws to hold it in.

Set the Housecode to the same letter as for the controller you will be using.

Set the Unit Code to a unique number (one that isn't used by another X-10 module).

The Wall Receptacle Module is now ready to receive commands from your controller.

Make a note of the Unit Code and Housecode settings.

Replace the cover plate with the new one provided.

Turn on the power at the circuit breaker or fuse box.

That's all there is to it. To test the module:

Plug an appliance into the controlled (top) outlet.

At the controller, press the Unit Code number and ON or OFF.

Make sure the appliance you want to control is always left on. It is now controlled by the Wall Receptacle Module.

Another feature of X-10 modules is their "local control" capability. If you are standing next to your appliance and don't want to walk to the controller, you can simply turn the appliance's switch off and then on again, and the appliance will turn on.

Thermostat Set-back Controller

The Thermostat Set-back Controller works with any type of household thermostat to regulate the heating and air conditioning in your home. This can dramatically reduce your utility bills, while maintaining a comfortable environment.

The Thermostat Set-back Controller works in conjunction with an Appliance Module or Wall Receptacle Module and a Mini Timer or Home Automation Interface to regulate the temperature throughout the day. If you use a Telephone Transponder as a controller, you can also dial in from any Touchtone phone and adjust the temperature of your house.

The Thermostat Set-back Controller attaches to the wall directly below your existing thermostat. It "fools" the thermostat by adding a predetermined amount of "heat." The thermostat then thinks that the room is warmer than it actually is, and is set back 5, 10, or 15 degrees, depending on how you've set the Thermostat Set-back Controller. This stops the heater from coming on until the room reaches a lower temperature.

In the summer you can use the controller in the opposite way. You set your thermostat for the highest temperature you want the room to attain, and the additional heat supplied by the Thermostat Set-back Controller fools the thermostat into thinking it is warmer in the room than it actually is. This turns the air conditioning on, to cool the house off before you get home.

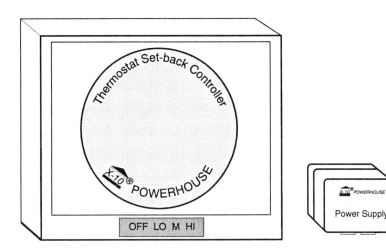

To mount and hook up the Thermostat Set-back Controller, you'll need wire strippers and a small, standard screwdriver.

Remove the front cover of the thermostat and use the provided mounting screws to attach the Thermostat Set-back Controller to the wall about one-quarter inch below the thermostat you want to control.

Adhesive-backed tape is included with the Thermostat Set-Back Controller, but should be used only for temporary mounting. Be sure to use the provided mounting screws to secure the controller permanently.

Replace the front cover.

Check for the location of the nearest electrical outlet. You'll be running the provided wire from the controller to the outlet. Plot a direct path that won't be obtrusive.

Attach the connected wire down the wall using wire staples or adhesive backing (included).

Run the wire along the floor or baseboard to the wall outlet.

If you need more wire to reach the outlet, be sure to use a similar gauge (thickness) of wire.

Strip one-half inch of plastic insulation from the end of the wire to expose two bare ends.

Connect the ends of the wire to the screw terminal on the Power Supply using the small, standard screwdriver.

If you wrap the wire starting on the left side of the screw terminal contact and wrap it over the top in a clockwise fashion, as you tighten the screw (clockwise) the wire will not come off as easily.

Plug the Power Supply into an Appliance Module, and then plug the module into a standard wall outlet or into the upper, controlled outlet of a Wall Receptacle Module.

The Thermostat Set-back Controller is now attached to the electrical system and able to receive commands.

Set the Housecode dial on the Appliance Module or Wall Receptacle Module to the same letter as for the controller you will be using.

Set the Unit Code to a number that is not being used by another module.

First you'll set it up to reduce heating during the winter while you're asleep or away from home.

Set your existing thermostat to the temperature you want your home to be while you're there (the warmest temperature).

The heater will maintain this temperature when the Appliance Module or Wall Receptacle Module is not off.

Set the Selector Switch on the bottom front of the Thermostat Set-back Controller to the desired amount of temperature reduction.

Lo (low) reduces the room temperature by approximately 5° Fahrenheit.

M (medium) reduces the room temperature by approximately 10° Fahrenheit.

Hi (high) reduces the room temperature by approximately 15° Fahrenheit.

The Lo, M, or Hi setting will reduce the temperature of the house by this amount when the Appliance Module or Wall Receptacle Module is on. Suppose you set your thermostat to 70° F and then set the Thermostat Set-back Controller to M (medium, or 10°). When the

Appliance Module or Wall Receptacle Module is on, the house temperature will remain at 60° F. When the Appliance Module or Wall Receptacle Module is off, the temperature will rise to the thermostat setting of 70°. In this scenario, you would set your Mini Timer or Home Automation Interface to turn the module on while you are away and turn it off just before you get home.

For air conditioning, you take a different approach.

Set the thermostat to the warmest temperature you want your house to reach while you are not there.

For example, 80° F might be a reasonable maximum temperature.

Set the Selector Switch on the bottom front of the Thermostat Set-back Controller to the desired amount of temperature reduction.

Lo (low) will reduce the room temperature by approximately 5° Fahrenheit.

M (medium) will reduce the room temperature by approximately 10° Fahrenheit.

Hi (high) will reduce the room temperature by approximately 15° Fahrenheit.

The Lo, M, or Hi setting will reduce the temperature of the house by this amount when the Appliance Module or Wall Receptacle Module is on. Suppose you set your thermostat to 80° F and then set the Thermostat Set-back Controller to M (medium, or 10°). When the Appliance Module or Wall Receptacle is on, the house temperature will be 70° F. When the Appliance Module or Wall Receptacle Module is off, the temperature will rise to the thermostat setting of 80°. In this scenario, you would set your Mini Timer or Home Automation Interface to turn the module on while you are at home and turn it off just before you leave.

Heavy-Duty Appliance Module

Not all appliances run on 110 volts. Many air conditioners, pool pumps and pool sweeps, and hot-water heaters require 220 volts of electricity. X-10 provides two different modules that operate almost identically to the 110-volt version, except that they allow for the additional capacity and load limits of these larger appliances.

The Heavy-Duty Appliance Modules work only with single- or split-phase systems. They do not work with three-phase wiring (which is sometimes found in apartment buildings).

If you're not sure which type of system you have, please consult a qualified electrician. You don't want to mess around with 220-volt systems.

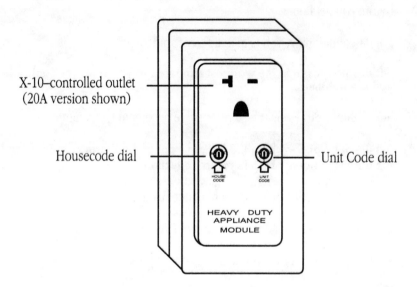

To select the appropriate Heavy-Duty Appliance Module, first determine the amperage rating of your appliance. If it is 15A or less, use the 15A Heavy-Duty Appliance Module. If it is 20A or less, use the 20A Heavy-Duty Appliance Module. This information should be somewhere on the appliance or in its manual. You should also note the type of plug on the appliance to make sure it matches the module.

Both modules are designed for single- or split-phase 110/220-volt systems. This is the most common wiring system found in houses. These modules do not work on three-phase systems.

Set the Housecode to the same letter you've set on the controller.

The Heavy-Duty Appliance Module is now ready to receive commands sent by your controller.

Set the Unit Code to a unique number (one that isn't being used by any other module).

Note that you *can* use the same Unit Code for more than one module. But be aware that modules using the same Unit Code will respond to the same commands, turning the devices on and off simultaneously.

Plug the appliance into the outlet on the front of the module.

Plug the module into a nonswitched 220-volt outlet (one that isn't controlled by a wall switch).

Bingo!

Test the module with your controller. Make sure the appliance's local switch is on. The appliance will now be controlled through the Appliance Module.

Press ON for the Unit Code that you assigned.

On comes the appliance.

The Heavy-Duty Appliance Module does not respond to the ALL LIGHTS ON command, but it does respond to the ALL UNITS OFF command. If you don't want your 220-volt appliance to be turned off when an ALL UNITS OFF command is set, you may want to put it on a separate Housecode. Then, for example, when you go to bed at night, you can press ALL UNITS OFF to shut off all the lights and appliance Modules on one Housecode without turning off your air conditioning or other 220-volt appliance.

Heavy-Duty Appliance Modules don't have a local control feature. You must use a controller to operate the attached appliance.

Universal Module

X-10 and other manufacturers have created modules to handle lights and appliances that run on 110 or 220 volts. There are many other devices that run on low voltages that don't plug into a standard outlet. For these devices, X-10 has created a special module, the Universal Module. The Universal Module has two screw terminal contacts on the bottom front of the module. Where a Light or Appliance Module turns off the device that is plugged into it, the Universal Module opens and closes the circuit between these two contacts, thus turning on or off any device that is properly attached to them. The CONTINUOUS/MOMENTARY slide switch allows you to tell the Universal Module whether to keep the contacts closed until it receives an OFF command or to close them for half a second and then reopen them.

The Universal Module also contains a **sounder** that beeps when an ON command is sent to its Housecode and Unit Code. The sounder can be used regardless of whether you use the screw terminal contacts to control a low-voltage device. This is useful if you would like to be audibly alerted when the contacts are closed and your low-voltage device is on.

In fact, this module can be used only as sounder for other modules. For example, you might want to know whether an Appliance Module in another room is on. If you set the Housecode and Unit Code of the Universal Module to the same Housecode and Unit Code as the Appliance Module, the sounder will sound either continuously or for three or four beeps, depending on how you set the MOMENTARY/CONTINUOUS slide switch.

The Universal Module is technically built to handle voltages up to 110-volt, 15A loads (500 watts for lights) and 1/3 horsepower inductive for motors, but the contacts are exposed and not insulated from being touched. Therefore, it is recommended that the Universal Module be used only for voltages of up to 30 volts.

 If voltage higher than 30 volts AC is switched through the Universal Module, the module must be placed in a locked, inaccessible place. THIS IS CRITICAL FOR SAFETY!

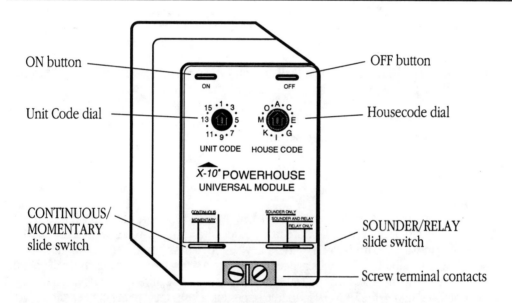

The Universal Module is relatively straightforward to set up, but provides many different options. The first example explains how to set up a low-voltage sprinkler, but you can apply the same techniques to many other low-voltage devices.

Set the Housecode to the same Housecode as for your controller. For now, use "A."

Set the Unit Code to a number that isn't being used by any other module.

You can think of the Universal Module as a simple switch. In fact, that's exactly how you hook it up, like a switch. A sprinkler system can be controlled by an **electric valve** powered by a low-voltage power source (a 24-volt power supply, in this example). When both contacts on the electric valve are connected to both wires on the power supply, and then it is plugged in and turned on, the electric valve opens and allows water to flow to the sprinklers. It wouldn't make much sense to install a fancy electric valve if you had to plug in and unplug the power supply every time you wanted to start and stop the sprinkler. That's where the Universal Module comes in.

This example uses an electric valve, a power supply (specified by the electric valve's manufacturer), the appropriate-gauge wire, and a Universal Module. Since there are many different types of electric valves, you'll need to check the specific requirements of the valve, but the concepts are the same. Make sure you use the recommended type of wire and power supply.

Run one wire from the power supply (either wire is fine) directly to one contact on the electric valve.

Run the second wire from the power supply to either of the screw terminals on the Universal Module.

Run a third wire from the unoccupied screw terminal on the Universal Module to the electric valve, and connect it to the unused connection.

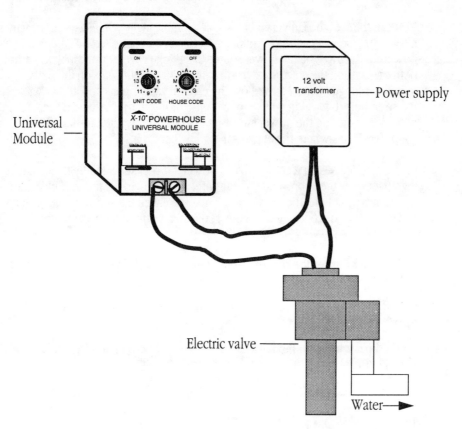

Universal Module

Power supply

Electric valve

Water→

The Universal Module now has the ability to open and close the circuit between its two contacts, and thus connect and disconnect one of the wires to the electric valve. That's all there is to wiring the Universal Module.

The slide switches give you a number of different options.

Move the CONTINUOUS/MOMENTARY slide switch to the CONTINUOUS position.

When the Universal Module is set to CONTINUOUS and receives an ON command, it closes the circuit between its two contacts (and turns on the low-voltage device) and keeps the circuit closed (on) until it receives an OFF command. If the slide switch is set to MOMENTARY and the Universal Module receives an ON command, it closes the contacts (and turns on the low-voltage device) for about one-half second, then it opens the contacts (turns off the device) and leaves them open.

For the sprinkler application, you'll use the CONTINUOUS setting (unless you are in a drought area and only want to water for half a second).

The Universal Module has a built-in sounder that can beep in conjunction with the closing of the contacts, by itself, or not at all. In some applications, you may want to hear a tone when the Universal Module receives an ON command. If the CONTINUOUS/MOMENTARY slide switch is set to CONTINUOUS, the sounder beeps constantly until an OFF command is received. With the MOMENTARY setting, the sounder beeps three or four times and stops.

It would be annoying to have the sounder beeping the entire time the sprinkler is on, so you'll turn it off.

Set the SOUNDER/RELAY slide switch to RELAY ONLY.

Now you'll give it a try.

Plug the power supply and the Universal Module into their respective electrical outlets.

The Universal Module has an ON button and an OFF button on the top for testing and local control. Try these first.

Push the ON button on the Universal Module.

The electric valve will open and the sprinkler will sprinkle.

Press the OFF button on the Universal Module.

The electric valve will close and the sprinkler will stop. Now test the Universal Module with your controller. As you can imagine, this setup works particularly well with controllers such as the Mini Timer and the Home Automation Interface. With these controllers, you can set up timed events to happen at certain times of the day, or in the case of the Home Automation Interface, a unique schedule for each day of the week. Try it with your controller.

From a controller with the same Housecode, push the Unit Code and ON.

On come the sprinklers.

Since X-10 is an **open-loop system,** meaning that there is no confirmation that a command has been received, sometimes it's a good idea to send commands a second time. In the case of a sprinkler system, this is a very good idea. If your Mini Timer sends the OFF command to your Universal Module to shut off your sprinklers, and for some reason the command isn't received (cosmic rays, perhaps?), you could have a swamp in your yard before you know it. An easy way to avoid this is to send the OFF command and then send another OFF command in a few minutes. If the module has been turned off, the second command won't do a thing, but if the module is still on, it could save you a lot of trouble.

The Universal Module can also be used to add extra features to your electric garage-door opener.

A typical electric garage-door opener uses a device called a **momentary switch** to signal the electric motor that opens and closes the door. This momentary switch probably looks and acts much like the push button for a doorbell.

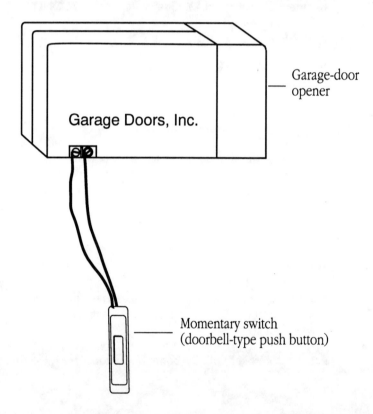

In this example, you'll use the Universal Module as an additional momentary switch so you can also use X-10 controllers to open or close the door. If your present garage door doesn't have a remote control, you can add this feature by using a Universal Module and a Wireless Remote Control. Even if your present garage-door opener has a remote control, by using X-10 modules and a Wireless Remote Control, you can consolidate your X-10 system and garage-door opener to work with one remote control. Then when you pull into your driveway, you can use an X-10 Wireless Remote Control to open or close your garage door, turn on your lights, and control any appropriate appliances.

Different garage-door openers have specific requirements, so be sure to read your manual to make sure that the following procedure will work with your door. Most modern garage doors have a safety feature that does not allow the door to close if an object is in its way. Using the Universal Module as described below should not disable any safety features, but be sure to test the door thoroughly to make sure all safety features are working after you add X-10 features.

Let's get started.

Locate the push-button switch that you presently use to open and close your garage door.

It has two wires that are connected when you push the button and disconnected when you release the button. You'll be using the Universal Module as an additional switch for the garage-door opener.

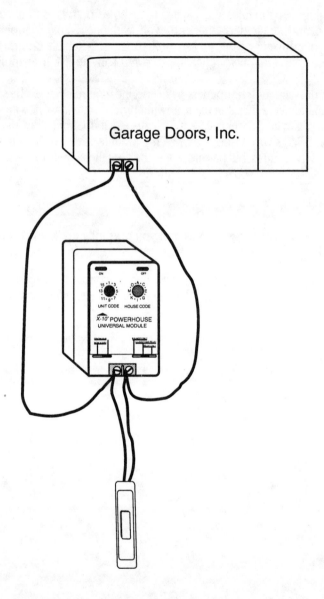

Disconnect the wires from the present push button.

If you need to cut the wires, be sure to leave enough to make a connection.

Attach the two wires that went to the push button to the screw terminals on the Universal Module.

You'll probably need to extend the wires so they can reach the Universal Module (which must be plugged into an electrical outlet).

Now you'll reattach the push button, so that you can use it for local control.

Run one wire from each of the screw terminal contacts to the connection on the push button. Do not disconnect the wires that are already attached to the screw terminal contacts.

Each terminal on the Universal Module will have two wires connected—one to the garage door and one to the push button.

Now you'll set the slide switches.

You'll use the MOMENTARY/CONTINUOUS slide switch to make the Universal Module simulate the pushing of a button.

Move the MOMENTARY/CONTINUOUS slide switch to MOMENTARY.

It might be helpful to have a warning when the garage door is being opened or closed. You'll use the SOUNDER/RELAY slide switch to turn on the sounder.

Move the SOUNDER/RELAY slide switch to SOUNDER & RELAY.

Now the sounder will beep three or four times every time the Universal Module hears an ON command and closes the circuit between the screw terminal contacts.

The circuit can now be closed using either the push button or the Universal Module. Give it a try.

Push the Universal Module's Unit Code and ON from any controller set to the same Housecode.

The garage door will open or close.

Neat, huh?

Push the original garage-door opener push button.

It should work just as it did before.

5 Home Security

Overview

In the previous chapters, you've seen the flexibility of X-10 technology and the variety of applications in which it's used. Another important application of home-control technology is home security. You've already seen how you can use controllers and light modules to light your house manually when you hear a suspicious noise. By installing sensors that sense the opening of a door or window or detect the presence of an intruder, you can achieve the same results—*automatically.*

First we'll explain some general terms, devices, and strategies common to any home security system.

Then we'll describe the Powerflash Interface, a device you can use to implement a simple security system with your existing X-10 system, or to integrate an existing alarm system with X-10 modules.

Most controllers and modules use the existing electrical wires in your house to send commands back and forth. In a security system installation, electrical outlets may not be conveniently placed near areas you want to protect. We'll describe the Supervised Wireless Security System, a complete security system that uses **radio frequencies** to communicate between **sensors** and the Base Receiver. This system takes the wiring hassles out of home security.

At the end of the chapter, you'll find answers to commonly asked questions and a comparison of the components that are available.

After reading this chapter, you should understand the applications for each of the components and be able to use them to design a professional-quality security system that is integrated with the other X-10 modules and controllers in your home.

Home Security Basics

All home security systems are made up of three components: sensors that detect a change (such as the opening of a window), **receivers** that monitor the sensors and initiate actions based on their status, and **alarms** that create a sound or otherwise alert you to the change in status of the sensor.

There are many different types of sensors, most of which can be integrated into an X-10 security system. One of the most common is the **magnetic contact switch.** This switch makes a reliable connection between two wires that make up a **circuit,** or circle of wire. You can imagine the problems that might occur if you attempted to get two wires to meet and separate consistently if the bare ends were exposed. It might work once or twice, but would not be reliable.

A magnetic contact switch makes this connection much more reliable—it attaches and detaches two wires via magnets. Instead of moving the wires together and apart, the magnetic contact switch uses a magnet on one end and a magnetically sensitive switch on the other. As the magnet approaches the two contacts on the switch, they are pulled together and the circuit is complete. This type of magnetic contact switch is called a **normally closed** contact switch. In its normal state, with a magnet acting on it is closed, and the wires are together.

Normally Closed Magnetic Contact Switch

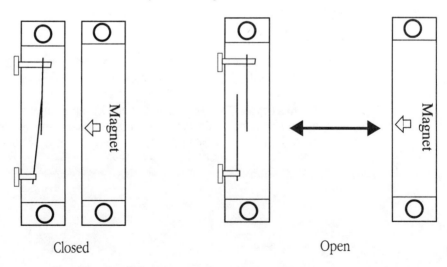

Closed   Open

When the magnet is moved away—for example, when a door or window is opened—the circuit opens and the wires are disconnected.

In cases where you generally want the circuit to be disconnected, with the wires unattached, you would use a **normally open** switch. When the magnet moves away from this type of switch, the contacts are pushed together and the circuit is closed.

Normally Open Magnetic Contact Switch

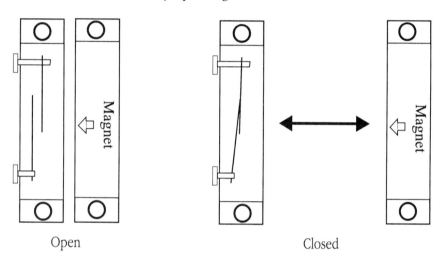

Open   Closed

Although either switch can be used with an X-10 system, there are advantages to a normally closed circuit. If an intruder notices your security system and cuts the wires of a normally open circuit, it will be disabled. With a normally closed system, however, cutting the wires will open the circuit, and the alarm will sound. Normally closed magnetic contact switches are the standard switches used for home security, and are included with most X-10 components.

By using a magnet to open or close the switch, you can place both halves of the circuit on one side of a window or door. The wires can be attached to the door or window frame, and the magnet can be attached to the door or window itself. This eliminates the need for long wires that allow for the opening and closing of the door or window. These switches also tend to be quite reliable and can stand up to the constant opening and closing with no noticeable wear.

Other Switches

Using a magnetic contact switch to monitor the status (open or closed) of a door or window is the most common application in home security. Many other devices are available that can open or close a circuit and set off an alarm or other signal.

Pressure mats are basically large, flat switches that sense pressure or weight. They are often used under a rug or carpet to detect someone walking into a room. Usually these switches are the normally open type.

Mercury switches are small, glass tubes with two exposed contacts and a drop of mercury that is capable of carrying electricity. Depending on the angle of the glass tube, the drop of mercury can be away from the contacts (circuit open) or against them (circuit closed). The mercury drop can carry electricity and completes the circuit. You could attach a mercury switch to a stereo component, for example. If someone lifts the component and the liquid mercury moves away from the contacts, the circuit will open and an alarm could sound.

Moisture sensors usually complete a normally open circuit. They can be used to detect water leakage or flooding.

Glass-breakage detectors contain a vibration sensor that can open a closed circuit using the vibration of a window being shattered.

Foil tape can form a vulnerable part of a normally closed circuit that will be broken (opened) when glass it's attached to shatters.

Once you understand the basic concepts of normally open and normally closed circuits, you'll be able to use a variety of different switches with the Powerflash Interface and the Supervised Home Security System. In the examples that follow, you'll learn how to implement magnetic contact switches, but consider using these other types of switches to enhance your system.

Powerflash Burglar Alarm Interface

Although the Powerflash Burglar Alarm Interface looks somewhat like a Lamp or Appliance Module, it's actually a special form of controller. Remember that a controller *sends* commands to modules and a module *receives* the commands and acts on them. The Powerflash Burglar Alarm Interface sends X-10 commands when it senses a connection between its two electrical contacts.

Any of the switch types described in the previous sections can be used between the electrical contacts of the PowerFlash Interface to detect an intruder. When the normally open switch closes, the circuit is completed and the Powerflash Interface sends the appropriate X-10 command. A MODE slide switch gives three different options for the commands that are sent.

The Powerflash Interface can also be used to connect an existing burglar alarm system with your X-10 modules to add more features to the overall system.

Many alarm systems use a low-voltage signal to trip warning devices. Using the INPUT slide switch, the Powerflash Interface can also detect these low-voltage signals (6 to 18 volts) and send X-10 commands. For example, by connecting the Powerflash Interface to your present alarm system, you could have all the lights in the house flash on and off, in addition to the alarm's normal signal. Individual alarms vary greatly, so consult your installer or the manual that came with your alarm system. We'll describe how to use the Powerflash Interface as a separate system to detect closure in a normally open circuit.

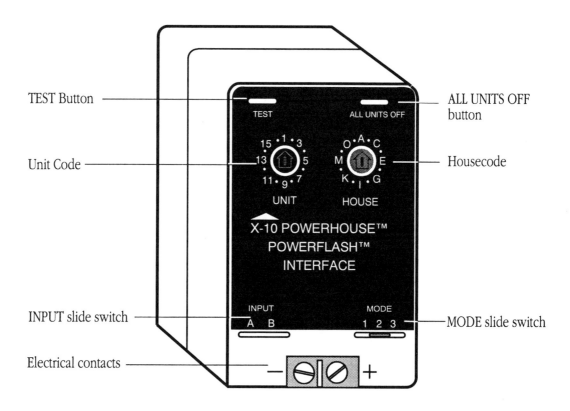

Powerflash Burglar Alarm Interface

The Powerflash Interface's different modes send different commands through the electrical wiring of your house to activate X-10 modules. The modes have the following functions:

Mode 1 When the Powerflash Interface detects a break in the circuit, it sends the ALL LIGHTS ON command to all Lamp Modules and Wall Switch Modules that have the same Housecode as the interface. It also sends an ON command to all Appliance Modules with the same Housecode and Unit Codes as the Interface. When the alarm condition is deactivated, all Lamp and Wall Switch Modules are left on, but modules with the same Unit Code as the Powerflash Interface will be turned off.

Mode 2 This mode makes the Powerflash flash all lights connected to Lamp Modules or Wall Switch Modules. When the alarm is deactivated, all Lamp and Wall Switch Modules will be left on, but Appliance Modules will be left off.

Mode 3 This setting makes the Powerflash turn on all Lamp and Wall Switch Modules that are set to the same Housecode as the Powerflash when the contacts are closed. When the alarm is deactivated and the contacts are open, the modules are turned off.

INPUT Slide Switch

The Powerflash Interface can be triggered by two types of changes in the circuit that connects the two electrical contacts. You'll choose INPUT A or INPUT B on the slide switch depending on the type of device you are connecting. The simplest application uses INPUT B to sense the closing of a normally open circuit.

INPUT A sets the Powerflash to sense a low-voltage signal, such as the signal sent by many conventional burglar alarm systems, a photocell, and other devices.

INPUT B sets the Powerflash to detect the opening of a normally closed circuit.

 Do not connect 120 volts AC directly to the electrical contacts. These contacts are rated at a maximum of 18 volts. If you connect higher voltages, severe damage and injury could result.

Now that you know the different options, you're ready to set up the Powerflash Interface.

Set the Housecode dial on the Powerflash to the same letter as on the modules you want to control.

When the Powerflash sends commands, your modules will receive them. You can set individual Housecodes for each group of modules as your system gets more sophisticated. For now, just use Housecode A.

Set the INPUT slide switch to B.

This sets the Powerflash to sense a change in a normally open circuit.

Set the MODE slide switch to 1.

This setting makes the Powerflash turn on all Lamp Modules and Wall Switch Modules that are set to the same Housecode as the Powerflash. The lights will remain on after the alarm is reset. Refer to the mode descriptions on the previous page if you'd like a different action to occur. Note that any Lamp Module set to the same Unit Code as the Powerflash Interface will turn off when the circuit closes.

Connect the electrical terminals to a normally open magnetic contact switch (purchased separately) using 18-gauge wire.

The Powerflash is compatible with normally open switches only when the mode is set to B. If you plan to use it to connect to an existing alarm, be sure to match the polarity (+ and -) with your alarm system.

Press the TEST button.

All Lamp and Wall Switch Modules that are set to the same Housecode will go on, as will Appliance Modules that are set to the same Housecode and Unit Code. This test ensures that the Powerflash is working correctly.

Press ALL OFF on the Powerflash Interface to turn the modules off again.

Close the normally open magnetic contact switch by moving the magnet away from it (by opening a door or window, if that is how you have connected it).

The results should be the same as pushing the TEST button. Experiment with different mode settings to see their effects. Remember, any normally open switch can be connected to the Powerflash.

Powerflash Interface Applications

The Powerflash Interface is very versatile and can be triggered by the low voltages (6 to 18 volts AC or DC) that are commonly produced by photocells, microphones, or virtually anything that sends a low voltage or closes a normally open circuit.

In Mode 1 and Mode 3, the Powerflash can send an ON command to an Appliance Module, which could turn on a stereo when you open a door, for example.

Supervised Home Security System

After exploring the options of the Powerflash Interface, you might consider buying a number of them to monitor your whole house. As your system expands, a few limitations become apparent. If you only have one door attached to the system and the alarm goes off, it's pretty clear what happened. If you have eight doors and windows connected, you may not be able to tell whether the problem is in the basement or in the attic. If a Powerflash Interface breaks or a magnetic contact switch goes bad, you wouldn't know without physically checking the system periodically.

Using the existing electrical system in your house is very convenient for controlling lamps and appliances, but can be inconvenient when you want to protect doors and windows. Using radio frequencies, commands can be sent between controllers and modules without any wires at all. X-10 has an entire product line of door/window sensors, plug-in base units, motion detectors, and sirens that can be combined to make a complete home security system.

The X-10 Protector Plus Supervised Home Security System combines a number of sophisticated features in a Base Receiver that monitors sensors and displays their status. It also allows for delays in entering and exiting, and other powerful features usually found only in more expensive systems.

The system is made up of several types of components: the plug-in Base Receiver, door/window radio transmitters (with magnetic contact switches or other types of switches), and hand-held remote control systems for arming and disarming the system. You can add other X-10–compatible products and more Door/Window Sensors to protect additional doors and windows. The plug-in Base Receiver can protect up to 16 doors or windows with individual sensors, and more than 16 when multiple doors and windows are attached to single sensors. In the security system, the Door/Window Sensors act as controllers (sending commands). The plug-in Base Receiver initially acts as a module (receiving the command), then transmits a response through the Power Line Carrier system to turn on or flash the lights, sound a siren, or take other action via standard X-10 commands.

The hand-held Remote Control can be used to arm and disarm the system remotely. Up to eight of these Remote Controls can be used by members of your family. The Remote Control can also be used for basic control of X-10–connected lights and appliances.

Base Receiver Placement

The placement of the Base Receiver is very important. Find a centrally located outlet that can't be seen from the main entrance of your home. This prevents an intruder from finding the source of the alarm and disconnecting it. The outlet that you choose should not be connected to a wall switch of any type. If the wall switch is inadvertently turned off, your alarm will also be turned off.

Because the Base Receiver's indicator lights are helpful in finding out which door or window is open and tracking down problems, it should be visible to family members. The Base Receiver also contains the alarm siren, so try to find an outlet from which the siren can be heard outside.

Remember that the Base Receiver should be:

- Hidden from the main entrance
- Connected to an unswitched outlet
- Visible to family members
- Audible outside the house

Setting Up the Base Receiver

Here's a diagram of the Base Receiver:

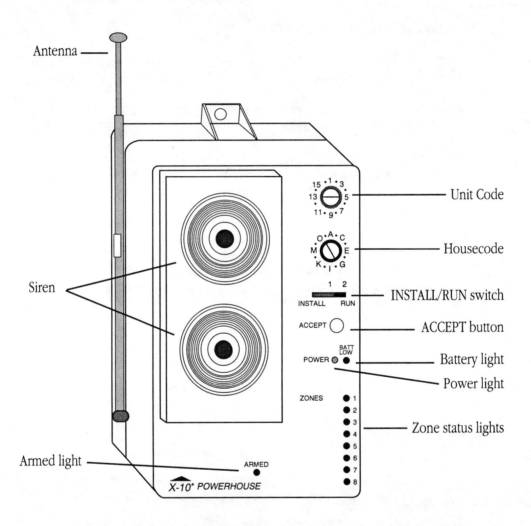

You may not be able to find an outlet that meets all of the requirements, but don't worry; just find the best location based on the criteria. Setting up the Base Receiver is easy. Here's what you do:

Set the Housecode on the Base Receiver to the same letter as on the rest of the modules you'd like to include in your security system. If you don't have any other modules, set the Housecode to A.

As indicated in the overview, the Base Receiver can turn lights and appliances on and off to deter intruders. Setting the Housecode to the same letter as other modules allows you to control the modules from the Base Receiver.

Set the Unit Code on the Base Receiver to an unused code (1 through 16). Set the INSTALL/RUN switch to INSTALL.

All the switches are now set for the initial installation.

Slide the battery cover off the top right corner of the Base Receiver and install a 9-volt alkaline battery. Replace the cover.

The battery maintains all the information in case there's a power outage or someone accidentally unplugs the Base Receiver. The system does not work while the power is out.

Plug the Base Receiver into the unswitched AC outlet you've chosen and extend the antenna.

The Base Receiver is ready to receive information from the Remote Controls and the Door/Window Sensors.

Setting Up the Remote Control

Here's a diagram of the Remote Control:

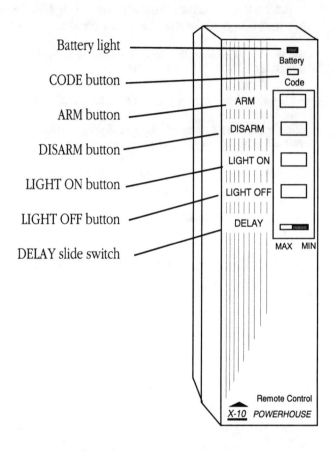

The full-size Remote Control can be hand-held or wall-mounted in an out-of-the-way, yet handy place. The Remote Control uses radio frequency signals to communicate with the Base Receiver. Radio frequencies allow you to control your system from inside your house, through walls, and even from outside a short distance away.

Slide the back cover off the Remote Control and snap in a 9-volt alkaline battery. Replace the cover.

Always use alkaline batteries in your Home Security System to ensure adequate voltage and long life. The Home Security System doesn't use code dials like other X-10 modules;

it sets its own addresses with an internal random code generator.

Make sure the Base Receiver's RUN/INSTALL slide switch is set to INSTALL.

Press the Remote Control's CODE button with the point of a pencil.

The CODE button is located just below the Battery light near the top of the Remote Control.

Press ARM on the Remote Control.

A tone will sound. The Base Receiver has identified the Remote Control and assigned it a unique code. Repeat these steps for any additional Remote Controls you may have, up to the maximum of eight per Base Receiver.

Installation for the Remote Control is complete.

Setting Up the Miniature Remote Control

The Miniature Remote Control provides most of the functions of the full-size Remote Control in a much smaller package. It is very convenient to keep on a key chain or in a purse. Installation is nearly identical to that of the full-size Remote Control.

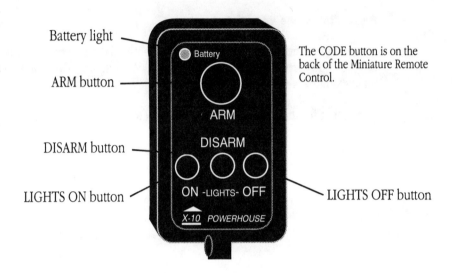

Make sure the Base Receiver's RUN/INSTALL slide switch is set to INSTALL.

Press the Remote Control's CODE button with the point of a pencil.

The CODE button is located on the back of the Miniature Remote Control. The Base Receiver assigns a unique code to the Miniature Remote, and it's ready to use.

Press ARM on the Remote Control.

Installation for the Miniature Remote Control is now complete.

Placement of the Door/Window Sensors

It's impractical to have a Door/Window Sensor on every door in your house. Who would want to break into your closets, anyway? Do place sensors on outside windows that are away from the street and hidden from view. Sliding glass doors and back doors are also a tempting target for intruders, as are garage doors. Walk around your house and think like a burglar. With the flexibility of the Door/Window Sensors and no need to run wires, you have a wide range of options.

You can mount Door/Window Sensors near the opening edge of any door or window. If you have pets or children, you might want to position the sensors high enough to be out of reach. You can also use 18-gauge wire to extend the distance between the magnetic contact switch and the Door/Window Sensor, if you'd like it out of sight. You can replace the existing wire by disconnecting it from the two screws located in the battery compartment.

Use double-sided tape (provided) to temporarily mount and test the Door/Window Sensor. When you're satisfied with the location and operation, be sure to use screws to make the installation more reliable.

For permanent installation:

Remove the battery cover from the back of the Door/Window Sensor.

Use the provided screws to attach the battery cover plate to the wall.

Don't screw the cover on too tightly, or you may break it. If the cover is too tight against the wall, it will be difficult to slide the Door/Window Sensor onto the mounted cover.

Slide the Door/Window Sensor down onto the mounted cover plate.

The magnetic contact switch should be mounted on the frame near the opening edge of the door or window. If you mount the switch on a metal door frame, make sure that the magnetic contact switch and the magnet (the half without screw terminals) are no more than 3/16 inch apart. On wood surfaces they can be 3/8 inch apart.

Be sure to install the magnet half of the switch (the one with no screw terminals) on the moving portion of the door or window, and the actual switch (with screw terminals) on the stationary portion.

Magnetic contact switches and magnets have arrows on them. Always make sure the two arrows are facing each other.

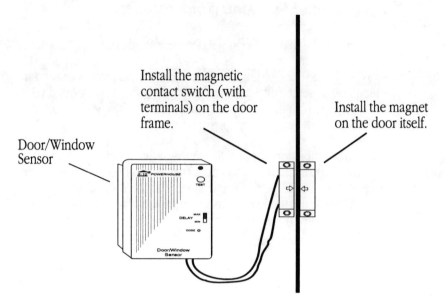

Mount the magnetic contact switch (with screw terminals) on the window or door frame (the stationary portion), and mount the magnet on the door or window itself.

Mounting the Door/Window Sensor near the top of the door will keep it from being kicked.

Connect the wires from the Door/Window Sensor to the screw terminals of the magnetic contact switch.

There is no order for connecting the wires; just ensure two good connections.

Check to make sure the two halves of the switch are well aligned. As you open and close the door or window, the light on the Door/Window Sensor should turn on.

Setting Up the Door/Window Sensor

You've already determined the best location for the Door/Window Sensor and temporarily mounted it. Here's a diagram of the sensor:

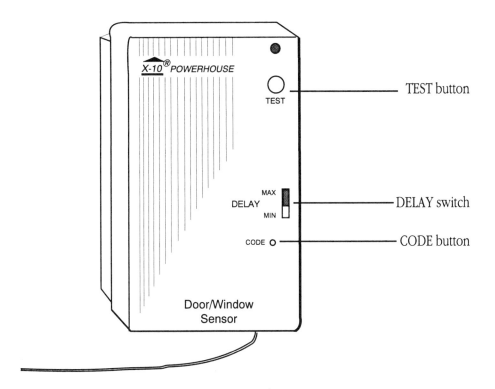

To set up the sensor:

Set the Base Receiver's RUN/INSTALL slide switch to INSTALL.

The Base Receiver is now ready to identify the Door/Window Sensors.

If you're using the Door/Window Sensor on a door:

Set the DELAY slide switch to MAX.

This setting will make the Door/Window Sensor send a command to the Base Receiver that tells it to give a pre-alarm beep and turn on the lamps and modules set to the same

Housecode and Unit Code as the Base Receiver. Thirty seconds later, if the system is armed in MAX mode, the alarm will sound and *all* lights connected to Lamps and Wall Switch Modules set to that Housecode will flash on and off.

If you're using the Door/Window Sensor on a window or door that should never be opened while the system is armed:

Set the DELAY slide switch to MIN.

When the door or window is opened, the alarm will trip instantly (even if the system is armed for delayed entry via the Remote Control—more on this in the Remote Control section).

Door/Window Sensors can operate with most types of normally open or normally closed switches. They come from the factory with a normally closed magnetic contact switch and are factory-set to normally closed. If you want to purchase your own sensor and it is normally open, there's a small slide switch in the battery compartment that changes the sensor to operate with a normally open switch. If you're using the supplied contact switch, leave it alone.

Press the CODE button on the Door/Window Sensor. It's located just below the DELAY slide switch on the front of the sensor.

Press the TEST button near the top right corner of the sensor.

The Base Receiver will emit a tone to indicate that it has accepted the sensor. The next unused zone indicator light will come on to indicate which zone the Door/Window Sensor was assigned.

Place the appropriate number sticker on the sensor to show its zone number.

It's important to keep track of which Door/Window Sensors are assigned to which zones. When an alarm goes off or there is a problem, the zone number will tell you which door or window is open.

Set the RUN/INSTALL slide switch on the Base Receiver back to RUN. The system is ready to arm.

First, we'll discuss different security options and then you'll finish arming the system.

Home Security Options

X-10 (USA) and other manufacturers have created a number of compatible additions to the basic Home Security System. Here are brief descriptions of a few of them.

Powerhorn

The Base Receiver has an 85-dB (**decibel**) built-in siren that will attract a great deal of attention, but may not be loud enough to truly deter a would-be burglar—especially in a large house. The Powerhorn is controlled by X-10 commands, so all you need is an electrical outlet. The Powerhorn is extremely loud—a piercing 110 dB! In fact, prolonged exposure to the Powerhorn siren could cause permanent ear damage.

The Powerhorn is very easy to set up and use.

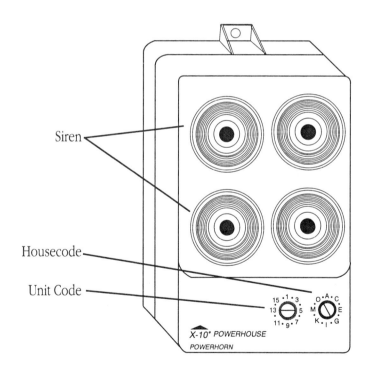

Set the Housecode and Unit Code on the Powerhorn to the same letter and number as for the security system's Base Receiver.

When the built-in siren on the Base Receiver sounds, the Powerhorn will also respond.

Find a nonswitched outlet (no wall switch control) away from the Base Receiver and plug the Powerhorn into the outlet.

The Powerhorn will receive commands from the Base Receiver and sound its piercing siren when the alarm is tripped. Next, we'll test the Powerhorn.

Stand away from the Powerhorn and trip the security system.

Lights connected to Lamp and Wall Switch Modules will flash on and off four or five times, and then the Powerhorn will sound. It will turn off a few seconds after the alarm resets or is disarmed.

If you want to activate the Powerhorn from the Powerflash Burglar Alarm Interface, simply set the Powerflash Interface to Mode 2 and set the Housecode and Unit Codes the same on each device. By using a Powerflash Interface and a Powerhorn, you can add a siren to your existing alarm with no special wiring.

Protector Plus Wireless Motion Detector

The Protector Plus Wireless Motion Detector is a specialized controller, sending X-10 commands to a Base Receiver that can then sound alarms or activate other X-10–controlled lights and appliances. The Wireless Motion Detector is—surprise—wireless and operates from a 9-volt battery. It can be set to trip immediately after detecting any motion or, to help prevent false alarms, it can be set to trip only if two movements or continuous movement is detected within a specified period of time.

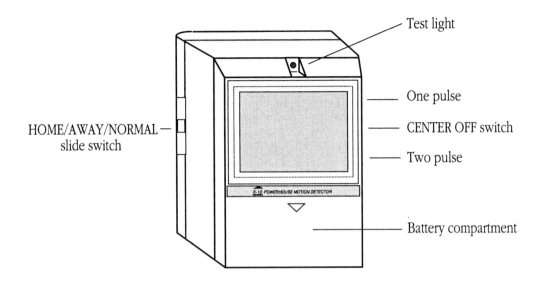

Placing the Wireless Motion Detector

Motion Detectors are most useful in central locations areas and points of heavy traffic. A single Motion Detector can provide coverage for most of a house if it is placed correctly. Find a main hallway or central entrance point for a group of rooms. This way you are protecting all of the rooms with one sensor.

Motion Detectors sense a change in temperature, including changes caused by cats, dogs, and air from heat ducts or air conditioning. If you have pets, the best way to avoid false alarms is to place the Motion Detector in a hallway that can be closed off to pets. As soon as any door is opened to the hallway, the alarm will trip.

You can make horizontal and vertical adjustments to the Wireless Motion Detector with the bracket that's included. You may be able to aim it high enough to avoid pets, but still detect an intruder. This will take some experimenting. The Wireless Motion Detector has a red test LED (light-emitting diode) that lights when motion is detected.

Use this LED to adjust the height and coverage area.

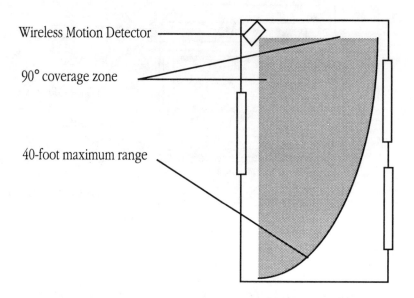

Installing the Wireless Motion Detector

Install a 9-volt alkaline battery in the battery compartment in the front panel of the Motion Detector.

Always use alkaline batteries in Home Security System units. They will give much better performance and require less frequent replacement. Some nonalkaline batteries may not work even when they are new.

Set the Base Receiver's INSTALL/RUN switch to INSTALL.

Now the Base Receiver is ready to identify the Wireless Motion Detector.

Press the CODE button on the back of the Motion Detector with the point of a pencil.

The unique code for the Motion Detector is established.

Press the TEST button on the back of the Motion Detector.

The Base Receiver will chime. The next available Zone Indicator LED will light up to indicate which zone number is being watched by the Wireless Motion Detector.

Each Wireless Motion Detector has its own zone, and the zone's status is indicated by an individual zone indicator light on the Base Receiver. Write down the zone number for each Motion Detector and keep the list near the Base Receiver.

If you don't hear a chime, press the CODE button, then press the TEST button again.

If you plan to use a Remote Control with the Home/Away feature to disarm your security system before you enter your house, you'll probably want to set the Motion Detector to the Home/Away position. If you want to use a Remote Control without Home/Away, set the Motion Detector to Normal.

To set the Motion Detector to trip the alarm if it detects any motion:

Set the CENTER OFF switch to 1 (a single movement).

To make the Motion Detector trip the alarm after it detects two movements or continuous movement:

Set the CENTER OFF switch to 2 (two movements).

If you don't want the Motion Detector to trip the alarm at all, set the CENTER OFF switch to the Normal position, in the center. The red test LED will still light when motion is detected. This is a good setting to use to test the system without setting off the ear-splitting siren.

Using the Supervised Security System

By now you should have placed and installed the Base Receiver, installed one or more Door/Window Sensors, and have the Remote Control in your hand. You may have installed a Wireless Motion Detector or a Powerhorn as well, but these are not required.

To test the system:

Set the INSTALL/RUN slide switch on the Base Receiver to RUN 2.

In Run 2 mode, the Base Receiver will chime when each door or window is opened instead of sounding the alarm. Much less annoying when you are testing the system.

Open each door or window in turn.

When each door or window is opened, the Base Receiver will chime and the appropriate zone indicator light turns on. When the door closes, the zone indicator light will turn off.

Arming the System in Instant Mode

Everything should be ready to go. Let's give it a try. (Don't do this late at night—the neighbors won't appreciate it.)

Set the INSTALL/RUN slide switch on the Base Receiver to Run 2.

Set the DELAY switch on the Remote Control to MIN.

The Miniature Remote Control is always set to MIN.

Press ARM on the Remote Control.

The Base Receiver beeps twice to acknowledge receiving the command.

Press TEST on the Door/Window Sensor.

This instantly sets off the alarm. The siren in the Base Receiver (and the Powerhorn, if installed) will sound, and any lights connected to Lamp Modules and Wall Switch Modules set to the same Housecode as the Base Receiver will flash on and off.

To stop the alarm:

Press DISARM.

The siren stops, and the lights stop flashing but remain on.

Test each of the Door/Window Sensors in this manner.

Arming the System in Delay Mode

The Delay mode gives you about one minute to leave the house before the system arms itself and about 30 seconds to disarm the system when you enter through an armed door (DELAY slide switch set to MAX).

Set the DELAY switch on the Remote Control to MAX.

Press ARM on the Remote Control.

Lamps connected to modules set to the same Housecode and Unit Code as the Base Receiver will turn on. Usually an entry light or a hall light is a good choice.

The Base Receiver chimes for about one minute while you leave the house. A beep signifies that the system is now armed, and Lamp or Wall Switch Modules with the same Housecode and Unit Code shut off.

Wait for at least a minute and open a door (DELAY slide switch set to MAX).

A pre-alarm beep sounds from the Base Receiver, and lights connected to Wall Switch and Lamp Modules with the same Housecode and Unit Code as the Base Receiver turn on. After 30 seconds, if the system is not disarmed, the alarm will sound and all lights connected to Lamp Modules and Wall Switch Modules with the same Housecode will flash.

If you open a window (DELAY slide switch set to MIN), the alarm trips instantly.

Press DISARM on the Remote Control.

The siren shuts off and the lights remain on.

Pressing any two adjacent buttons (LIGHT OFF and LIGHT ON or ARM and DISARM) simultaneously will sound the alarm instantly.

Using the Miniature Remote Control

The Miniature Remote Control performs the same functions as the Remote Control, but always activates the system in Instant mode. There is no entry or exit delay. It also has a Panic Alarm that you can set off by pressing ARM and DISARM at the same time.

The Miniature Remote Control will also perform LIGHT ON and LIGHT OFF functions for any module that is set to the Base Receiver's Housecode and Unit Code.

Trouble Alarm

If you try to arm the system and hear a continuous two-tone sound, there is a problem and the system will not arm. You can either correct the problem or ignore it by bypassing the sensor that is causing the problem.

To correct the problem:

Press DISARM and check the sensor located in the zone with the flashing indicator light.

To ignore the problem:

Press the ACCEPT button on the Base Receiver.

Press ARM on the Remote Control again.

The problem zone indicator will flash rapidly. The system will arm, but the problem zone is bypassed and thus not protected.

6 *Apple Macintosh*

Overview

Assuming you haven't skipped right to this chapter, you've already learned a lot about home automation. In fact, you already know enough to build a very functional system that will control lights, appliances, even a security system.

In this chapter, you'll learn how you can control a reasonably complex home-automation system with a Home Automation Interface and an Apple Macintosh computer.

Using the interface with a Macintosh computer gives you several advantages over systems that use the other controllers described in this book:

- You can see "icons," or pictures, of each module on the computer screen, so it's easy to remember what modules control which devices.
- You can control modules set with *any* Housecode and *any* Unit Code; this means that you can control up to 16 x 16 = 256 devices!
- You can "program" up to 128 different timed events.
- You can "program" each module to go on and off at specific times, and you can also program modules to go on and off only on certain days of the week.
- You can program Lamp Modules to dim to a certain intensity at a given time.

As you go through this chapter, you'll first learn how to attach the Home Automation Interface to your Macintosh. Then you'll learn how to create **module icons,** and how to use them to control devices both immediately and with timed events. Finally, you'll learn some advanced commands for displaying different floor plans on the screen, and for managing the interface.

When you're finished reading this chapter, you should be able to use your Macintosh to turn devices on and off both immediately and by using timed events. You'll also know how to create a customized floor plan for the Home Automation software, so you can more easily identify modules in your home.

This chapter assumes that you are familiar with the operation of the Macintosh computer: using menus, dragging, double-clicking, and so on. If these terms are not familiar to you, please consult your Macintosh owner's guide before continuing.

Setup

Before you try to use the interface with your computer, you'll need to ensure that the interface is functioning properly. Here's how to do that:

- Set the address of a Lamp Module to A1, attach the module to a lamp, and then plug the module into an electrical outlet.
- Next, plug the interface into a different electrical outlet. (Although you can use any electrical outlet in the house, choose one in the same room so you don't have to run around a lot.)
- Press the top of the rocker button labeled "1" on the interface to turn the lamp on. Once the lamp has turned on, press the bottom of the same button to turn the lamp off.
- Set the Lamp Module to A2, and repeat the preceding step with rocker button 2.
- Continue as outlined above, testing rocker buttons 3 through 8.

If you experience any problems, be sure you have set the Lamp Module correctly. If you still have problems, call X-10 (USA) Inc.'s Customer Service Department for help (the telephone number is given in Appendix C, page 173).

One last thing you should do before connecting the interface to your computer is to install a 9-volt battery in the battery compartment on the back of the interface. The battery will provide backup power to the interface when it is not plugged into an outlet or when the electricity is off. Without the battery installed, you'll have to reprogram every timed event whenever the interface is unplugged from an electrical outlet or the power goes off.

Connecting the Interface to Your Computer

Setting up the interface is straightforward. The first thing you need to do is attach it to your Macintosh computer. You do this using the serial cable that is included with the interface.

Locate the serial cable that is included with the interface. Then plug it into the back of the interface.

Once you've examined the connectors on both ends of the serial cable, it should be obvious which end goes into the interface.

Next, you'll connect the serial cable to the modem port of the Macintosh. To locate the modem port, look at the group of connectors, or **ports,** on the back of your Macintosh. Each port is designated by a different icon embossed on the plastic just above or below it. The modem port has an icon of a telephone next to it. Now that you've located the modem port, you're ready to connect the cable to the Macintosh.

Plug the other end of the serial cable into the modem port on the Macintosh.

The next step is to copy the X-10 software onto your hard disk.

Make a folder on your hard disk and name it "Home Automation" or something similar. When you are finished, insert the X-10 disk into your Macintosh computer's disk drive and copy all of the files on the disk into the new folder.

If you open the Home Automation folder you just created, here's what you should see:

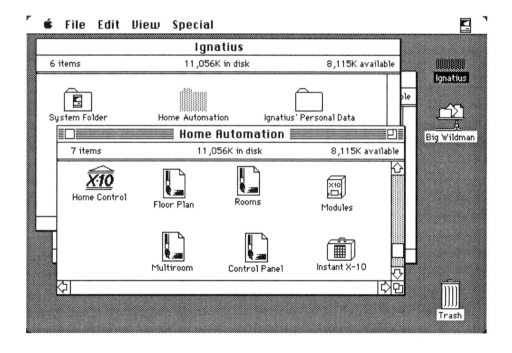

Now you should be ready to go.

Getting Started

The first thing you'll do is start the application.

Double-click the Home Control application to open it.

Because this is the first time you're using the interface, it contains no data, and you'll see the dialog box shown below:

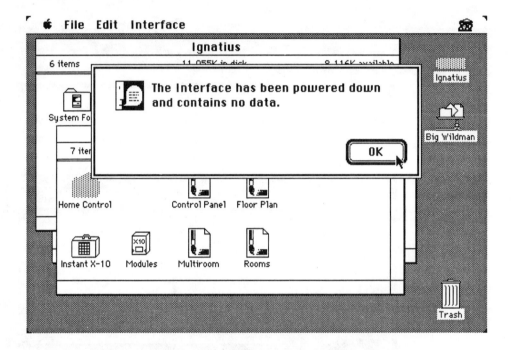

This, of course, is no life-threatening matter. It just means that you'll have to send information to the interface before you can use it to control things.

Click OK.

You can use the interface for timed events, similar to the Mini Timer described in Chapter 2. The interface has a clock that it uses to support timed events, and the first piece of information you need to give the interface is the time and the day of the week. Because your Macintosh knows this information, the X-10 software offers to set the interface clock to the same time as the Macintosh clock.

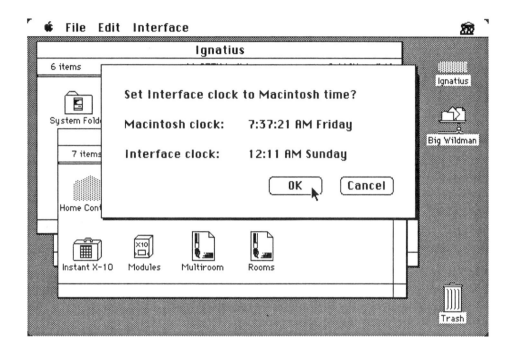

Click OK to set the interface clock.

Note that if the clock in your Macintosh is not set correctly, then the clock in the interface will not be set correctly either. If necessary, refer to the section "Setting the Interface Clock" later in this chapter.

Now you should be face to face with the main screen of the X-10 Home Automation software. The screen is divided into two areas: the **Control Area** and the **Module Map**. The Control Area is on the left. An overview of the controls is shown below.

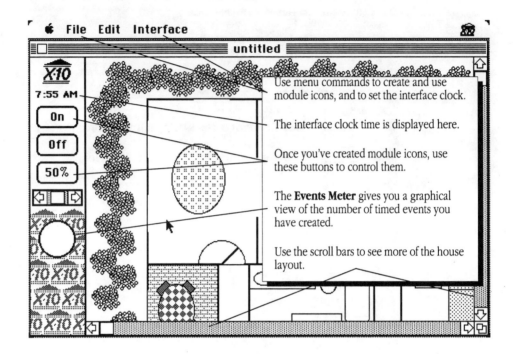

The Module Map is on the right. It consists of a MacPaint-style diagram, called the background, and a group of module icons. The default MacPaint background (the one you see now) represents the floor plan of a sample home. If you're like us, the floor plan doesn't look at all like your house. Don't worry—you can change it later.

You'll also create the other part of the Module Map—the module icons—later. Then you'll use the icons to control a lamp that you set up where you are working. Later, you can use what you've learned to control other devices: appliances, sprinklers, outdoor lighting, and so on. Before you start creating module icons, though, it's a good idea to test the interface to make sure it's functioning properly. You'll learn how to test the interface now.

Testing the Interface

The interface software includes a self-test you can run to ensure that it's operating correctly. Unfortunately, when you run the self-test, you erase all of the information in the interface; therefore, it's a good idea to run the self-test now, before you've created any timed events or stored any module information.

Choose Self Test from the Interface menu.

A dialog box appears, asking if you're sure you want to run the self-test, and warning you that doing so will erase all data in the interface.

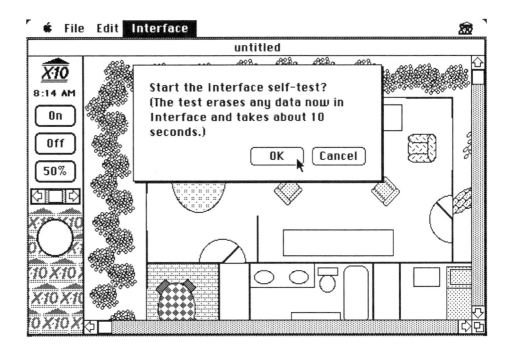

Click OK.

Next, a dialog box tells you that the test is running. If your interface has no problems, you'll see the dialog box shown below when the test is finished.

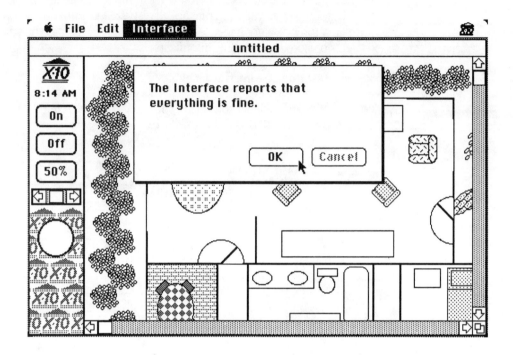

Click OK.

If you see a dialog box telling you that the interface is having problems, you should return the interface and exchange it for a new one.

Now you're ready to begin creating module icons and controlling devices.

Creating and Using a Module Icon

Like most Macintosh applications, the Home Automation application has a graphical user interface. To control devices, you create an **icon,** or picture, of the device, and then choose menu commands to set up and control the device.

You'll create a module icon now. For this simple example, you should first set a Lamp Module to Housecode C and Unit Code 5, attach it to a lamp, and plug the module into an electrical outlet in the same room as your Macintosh. That way, when you're doing the exercise that follows, you'll be able to verify that everything is working correctly as you go along. Please note that if you do not use Housecode C and Unit Code 5, the instructions that follow will not work for your module.

If you don't want to continue to use Housecode C for the lamps in your system, don't worry. You can reconfigure the Lamp Module once you've finished this exercise.

Now you're ready to create that module icon.

Choose New Module from the Edit menu.

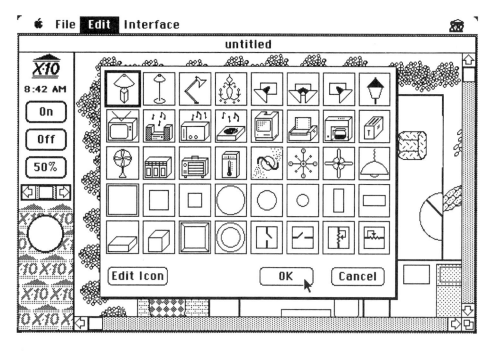

A dialog box appears, allowing you to choose the type of module icon you want to create. Each picture represents a different kind of device you can control. The top row contains icons of various types of lamps. The second row contains appliances, and the third row contains more appliances and lamps. The bottom two rows are for controlling alarm systems and for designing your own icons for other uses.

Creating and Using a Module Icon 101

The default, or preselected, icon is the one at the far left of the top row. You can tell it's selected because it has a heavy black box around it. To select a different icon, you would just click it. In this case, though, you're controlling a lamp, so you'll use the default icon.

Click OK.

The Module Setup dialog box appears. Here, you enter a name and set the Housecode and the Unit Code for the module. First, you'll enter a name for the module.

Type a name for the module.

Choose a unique name that will help you identify the module later, when the screen is full of modules.

Next, you'll set the Housecode. To change the Housecode, you just click in the box that contains it (see diagram below). This will advance the Housecode by one letter. You'll set the Housecode to C now.

Click the Housecode shown in the dialog box.

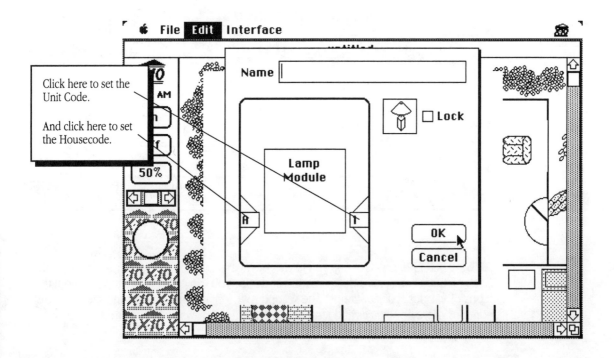

The Housecode changes to B. If it has not changed, you've clicked in the wrong area. Look again at the diagram, and make sure you click in the box that contains the A, in the lower left part of the dialog box. You want to set the Housecode to C, so…

Click the Housecode again to change from B to C.

You change the Unit Code the same way. You'll change it to 5 now.

Click the Unit Code four times.

Your screen should now look like the one below. If not, review the previous instructions and try again.

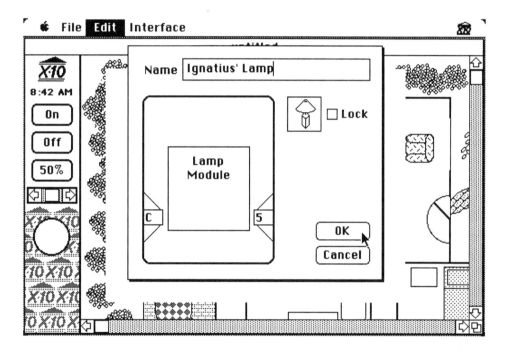

Now that you've set up the module, you're ready to actually do something with it.

Click OK.

Creating and Using a Module Icon   103

The dialog box closes, and you see the module icon on your screen. Notice that it's labeled with the name, Housecode, and Unit Code you selected. Also notice that it landed squarely on a chair in the living room. The first thing you'll do is move the module icon off the chair. If you're familiar with Macintosh, you've probably already guessed that to move a module icon, you just drag it.

Drag the module icon you just created a little to the right, between the two chairs shown in the diagram.

Your screen should now look like the one shown below.

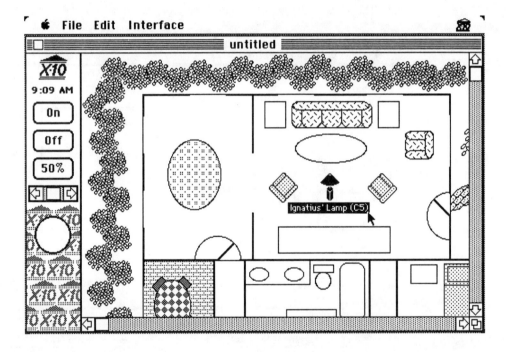

Now you're ready to control the lamp. There's one simple secret you need to know to control devices with the Home Automation software:

First you select a module icon. Then you choose commands to control that module.

You'll control the lamp now. First, though, you need to ensure that it is selected. A module icon is selected when it is highlighted (blackened) and its name appears below the icon. Remember that to select a module icon, you click it.

        Selected   Not selected

First things first: You'll turn the lamp on. To do this, look at the upper left portion of the screen. There are three buttons: On, Off, and 50%. By clicking the first button, you can turn the selected module on.

[On]   **Click On.**

The lamp in your room should turn on. If it doesn't turn on, check the module attached to it to ensure that the Housecode and Unit Code are set correctly (C5). Then make sure that the module icon is selected. Remember that when a module icon is selected, it's highlighted, and the module name, Housecode, and Unit Code appear below it (see diagram above).

Next you'll turn the lamp off.

[Off]   **Click Off.**

The light goes off. Pretty simple, eh? Hang on. Things get a little more complicated, but not much. Next, you're going to dim the lamp. To do this, make sure you've attached the lamp to a Lamp Module, and not to an Appliance Module. You cannot dim lamps that are attached to an Appliance Module.

[50%]   **Click the button labeled "50%."**

The lamp turns on and dims to half its original intensity.

You can control brightness with the sliding control that appears below the 50% button. You change the intensity by dragging the white elevator box to the right or the left. Dragging to the right increases brightness; dragging to the left decreases brightness. You can also move the box by clicking the arrows (see diagram below).

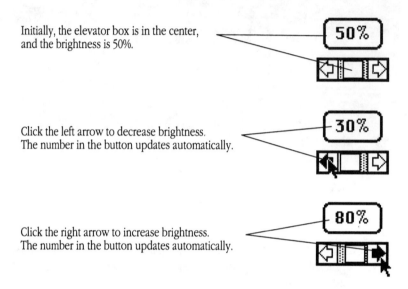

Initially, the elevator box is in the center, and the brightness is 50%.

Click the left arrow to decrease brightness. The number in the button updates automatically.

Click the right arrow to increase brightness. The number in the button updates automatically.

Now you've learned how to create a module icon, how to use it to turn lights and appliances on and off, and how to dim lamps. You can also use module icons to control devices that are attached to a Universal Module. For example, you could use the same procedure to create a module icon to control low-voltage lights or sprinklers.

Setting Up Timed Events

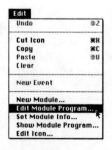

Now that you know how to control devices immediately, you're ready to learn how to create timed events. You do this by "programming" the module icons that you create. But don't worry—you don't have to be a programmer to figure it out.

You'll program your lamp module icon now.

Choose Edit Module Program from the Edit menu.

A dialog box appears. The parts of the dialog box are described briefly below.

Use the clock to set the time of the event.

Use the ON/OFF switch or the **Dimmer Control** to set the type of event you want to create.

Select Today or Tomorrow for an event you want to occur only once.

Or click Weekly to create an event that recurs throughout the week.

Then select the days on which you want the event to happen. Here, Tuesday and Thursday are selected.

Use the Events List scroll bar to scroll through the different events you've created for the module icon. (You haven't created any yet, so none are shown.)

The Edit Program Module dialog box allows you to create timed events. That is, it allows you to turn a device on and off at specific times, and to dim lamps attached to Lamp Modules or Wall Switch Modules at certain times. There are four things you must do to create a timed event:

- Create the event by clicking the New Event button or choosing New Event from the Edit menu.
- Select an event type: On, Off, or Dim.
- Set the time of day for the event.
- Choose a day of the week for the event: Today, Tomorrow, or Weekly (recurring).

You'll go through the steps to create a timed event to turn your lamp on. If you've been following along, the lamp is turned on and dimmed 50%. You should turn it off before creating a timed event to turn it on.

Close the dialog box by clicking its close box. Then select the Lamp Module you created and click OFF to turn the lamp off. Finally, choose Edit Module Program from the Edit menu.

Now you're ready to create your first event.

The first thing you need to do is create a new event.

Click New Event (or choose New Event from the Edit menu).

Next, you need to set the event type: On, Off, or Dim. Look at the box in the upper-left corner of the dialog box. The box resembles a light switch and works the same way. That is, you create an Off event by setting the switch to OFF, and an On event by setting the switch to ON. You change the switch setting by clicking the word ON or OFF.

Alternatively, you can create a Dim event by clicking in the Dimmer Control and moving the elevator box up (brighter) or down (dimmer). Note that the elevator box does not appear until you click in the Dimmer Control.

See the diagram below.

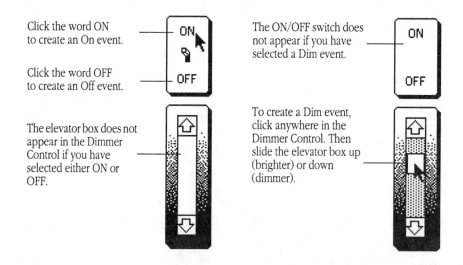

You want to create an On event.

Change the event type by clicking ON on the ON/OFF switch.

Next, you need to set the day of the week for the event. You do this by pressing one of the three buttons labeled Today, Tomorrow, and Weekly (see diagram below). Choose Today or Tomorrow if you want the event to occur only once. If you want the event to recur week after week, choose Weekly.

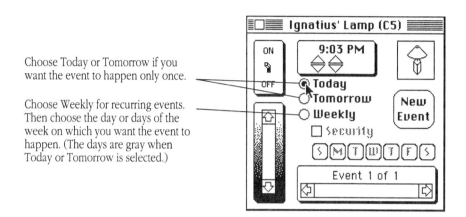

Choose Today or Tomorrow if you want the event to happen only once.

Choose Weekly for recurring events. Then choose the day or days of the week on which you want the event to happen. (The days are gray when Today or Tomorrow is selected.)

For this exercise, you just want the light to turn on today.

Click Today.

Next you need to set the event time. You'll use the arrows below the clock in the dialog box.

Because this is your first timed event, you're probably a little skeptical, so you'll want to actually see the lamp go on. To do this, you'll set the event for five minutes from the current time.

Set the clock to five minutes past the time shown in the Control Area by clicking the upper-right arrow shown in the diagram.

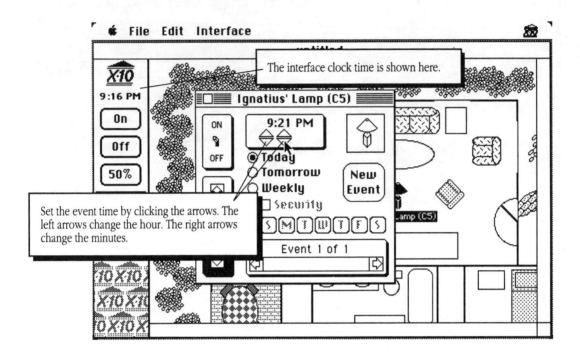

That's it! You'll need to know more about creating timed events, but that can wait until later.

Now you've created the event, but there's one more thing to do before the light will turn on. So far, you've created the event with the software that runs on your Macintosh, but remember that it's the *interface* that needs to communicate with the module connected to your lamp. Before your events can take place, you need to send them to the interface.

You'll do that now, but first you should check to see that the time you've set for the event is later than the current interface time shown in the Control Area. If not, click the arrow to move the clock forward a few minutes.

110   Apple Macintosh

Choose Send Events to Interface from the Interface menu.

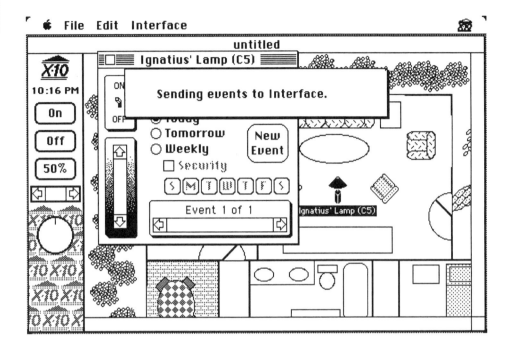

A dialog box appears briefly, telling you that the Home Automation software is sending the events you've created to the interface. Once the event has been sent, the dialog box will disappear, and all you need to do is wait five minutes for the clock in the Control Area to catch up to the time you set for the event. Then, your lamp should turn on, and you should be ready to continue.

If the time in the Control Area catches up to the time you set and the light *doesn't* turn on, check to make sure you've set the Lamp Module correctly. Also check that you've set the time correctly; specifically, make sure that you have correctly specified AM or PM. If you still can't solve the problem, see Appendix B, "Troubleshooting," on page 163.

Now you're ready to create a second event. This time, you'll create a recurring event—that is, one that happens every week at the same time. As always, the first thing you need to do is create the event.

Click New Event.

Notice that the text in the Event List changes to "Event 1 of 2." Now that you've created multiple events for the module icon, you need to know which one you are currently viewing. You'll learn how to switch between timed events for a module icon later in this section.

Next, you'll need to specify the event type. You'll create an event to turn the lamp on.

Click ON to set the event type.

Now you're ready to select the day of the week. This time, you want to create a recurring event.

Select Weekly by clicking the radio button to the left of it.

When you click Weekly, you also have to specify which days of the week you want the event to occur. You do this by clicking one or more of the buttons labeled S (for Sunday) through S (for Saturday). Note that if a button is highlighted (blackened), the event will occur on that day. If it isn't highlighted, the event will not occur on that day. If you accidently select a day, click it again to deselect it.

Set the event to happen every Monday, Wednesday, and Friday: Click M, W, and F.

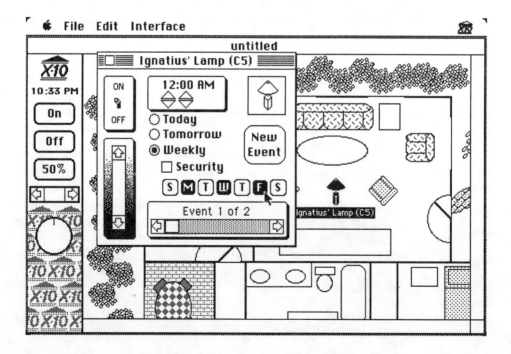

Next, you could set the time to any value. For now, leave it at the default: 12:00 AM.

That's it. Now you've got two events set for your lamp. You could set more events simply by following the procedures you've just learned.

Remember, once you've created more than one event for a module, you can view and modify the events by dragging the elevator box in the Event List at the bottom of the Edit Module Program dialog box (see diagram below).

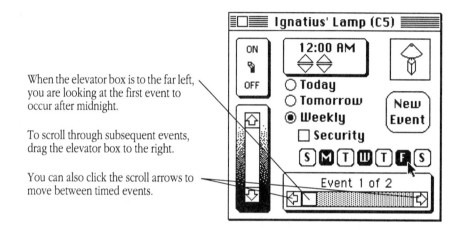

When the elevator box is to the far left, you are looking at the first event to occur after midnight.

To scroll through subsequent events, drag the elevator box to the right.

You can also click the scroll arrows to move between timed events.

Drag the elevator box in the Event List to the right to display the second event.

You should now see the first event you created, which will turn the lamp on for today only. Note that although you created this event first, it appears second in the Event List because it occurs later in the day.

Removing Events

From time to time, you'll certainly want to change the events you've created for a module. For example, you may want to change the time of an event, or even delete it entirely. Modifying an event is easy: You just scroll to the event and change whatever you want. Removing an event is also easy: As with any Macintosh application, you just "cut" it out. You'll cut your first event now.

Make sure that the first event you created is showing. Then choose Cut Event from the Edit menu.

The event vanishes, and you're left with only one event, which turns the lamp on at 12:00 AM each Monday, Wednesday, and Friday.

Reviewing Events

If you have created many events, it might be difficult to tell when the device will be on or off just by scrolling through the Event List. Luckily, there is an easier way. You can see a graphical representation of all the events for a module by using the Show Module Program command.

Choose Show Module Program from the Edit menu.

A dialog box appears, showing you the Event List for your lamp.

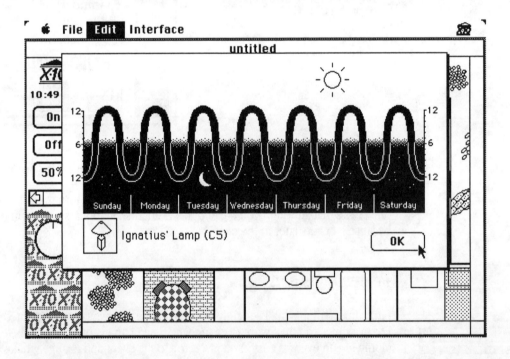

The dark line shows when the module is on. The diagram above shows that your lamp is always on. This may seem strange, but remember that you've created only *one* weekly event, and that event turns the lamp on. Therefore, the lamp is on all week.

Now you'll create an event to turn the lamp off in the morning, and see how that affects the module program.

Click OK. Then create a new event to turn the lamp off every Monday, Wednesday, and Friday at 6:00 AM.

If you have forgotten how to do this, review the instructions given on page 108.

Now you're ready to review the module program for the week.

Choose Show Module Program from the Edit menu.

Now your module program should look like the one below.

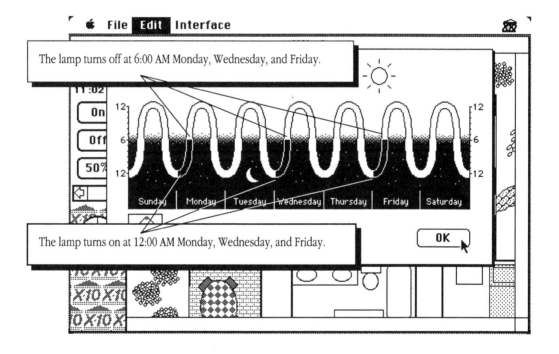

The Show Module Program feature is especially useful when you are reviewing programs for lamps, heaters, sprinkler heads, and so on. Imagine if you created a timed event to turn the sprinklers on, but forgot to create one to turn them off!

When you've finished reviewing the module program, click OK. Then close the Edit Module Program dialog box by clicking its close box.

Checking the Events Meter

Now you've learned how to create module icons and events, and how to send events to the interface. As you may already have guessed, the interface cannot hold an infinite number of events. In fact, it can store only 128 events. Once you've created many module icons, and several events for each icon, it might be difficult to keep track of how many events you've defined, and how many more you can still define.

The Home Automation software gives you a visual clue, an **Events Meter,** that always lets you know how much interface storage you have used.

The Events Meter is located in the Control Area (see diagram below).

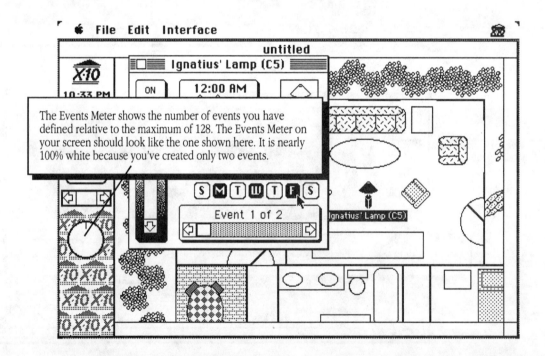

The Events Meter starts out 100% white, and becomes increasingly blackened as you create timed events and fill the interface memory.

Changing Module Setup Information

So far, you've learned how to create module icons, how to create timed events, how to modify events, and how to delete events. But just as you might want to change individual events from time to time, or even delete them entirely, you might also want to change module icon information or delete module icons. You'll learn how to do that now.

Let's say that you decide to set all Lamp Modules to Housecode A so you can turn them all off at once with a Mini Controller that also uses Housecode A. You can easily change the module setup with the Set Module Info command. First, though, you'll need to select the module icon for your lamp.

Make sure the lamp module icon you created is selected (highlighted).

Now you're ready to change the module information.

Choose Set Module Info from the Edit menu.

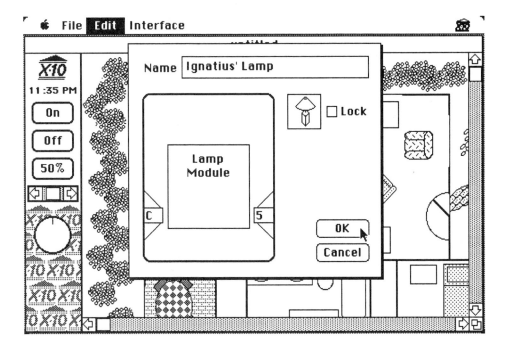

Setting Up Timed Events   117

A dialog box appears, allowing you to modify the Housecode and Unit Code for the module icon. Remember that to change the Housecode, you click the current Housecode to advance it by one. Similarly, you click the current Unit Code to advance it by one.

Practice changing the Housecode and Unit Code. When you are finished, return the settings as they were and click OK.

Setting the Base Housecode

Sometimes, when you're in a hurry, you may not want to turn your Macintosh on and run the Home Automation software just to turn a light on or an appliance off. To make life easier, there are eight buttons on the top of the interface that you can use to turn devices with Unit Codes 1 through 8 on or off.

At this point, you should be asking "OK, Unit Codes 1 through 8, but which Housecode?" You might guess that the interface always uses Housecode A. But you'd be wrong. In fact, the interface can use *any* Housecode. You just have to tell it which one you want to use. You tell the interface to use the rocker buttons with a different Housecode by changing the **Base Housecode.**

Right now, you've got one module icon set to Housecode C. Therefore, it makes good sense to change the Base Housecode of the interface to C. You'll do that now.

Choose Set Base Housecode from the Interface menu.

A dialog box appears, telling you that changing the Base Housecode will erase all events sent to the interface, and asking if you want to proceed.

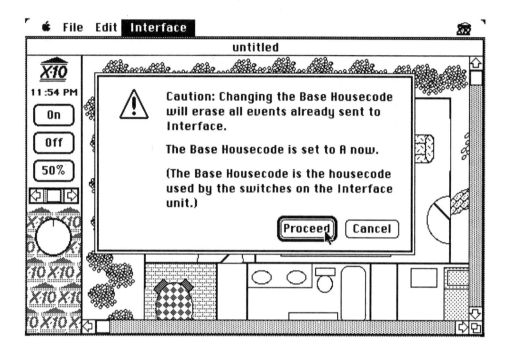

If you'd created many module icons and events, you might be hesitant about changing the Housecode. However, you have only one module icon and two events, so it'll produce no massive trauma.

Click Proceed.

Next, you see a dialog box that will allow you to change the Base Housecode.

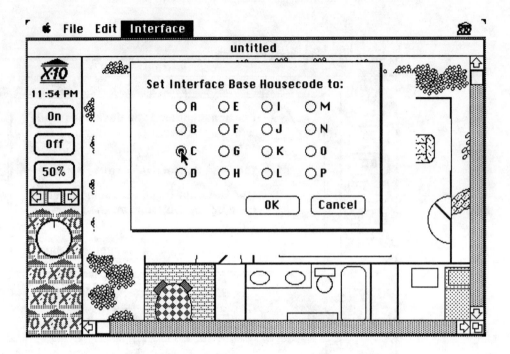

Here you can change the Base Housecode by clicking the radio button next to the appropriate letter.

Click the radio button next to the letter "C," as shown above. Then click OK.

Now you can use the rocker buttons on the top of the interface to control your lamp.

Press the bottom of the button labeled "5" on the top of the interface to turn your lamp off.

Pretty nifty.

Deleting Module Icons

You've learned how to create, modify, and delete events for module icons. But you've only learned to create and modify the module icons themselves. There are certainly going to be times when you want to delete module icons entirely, though, and it's about time you learned how.

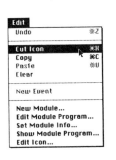

Make sure that your module icon is selected. Then choose Cut Icon from the Edit menu.

The icon disappears.

Remember, though, that the events you've created won't be erased from the interface until you choose the "Send Events to Interface" command.

If you decide that you really didn't want to delete the icon, it's easy to get it back.

Choose Paste from the Edit menu.

The icon reappears.

You've learned almost everything about creating module icons and events and controlling the interface with the Home Automation software. The rest of this chapter explains other useful commands you can use, and provides some tips and hints that should help you get the most out of your X-10 system.

Configuring and Customizing

Setting the Interface Clock

When you first used the interface, it contained no data, and you set the interface clock to the time of the Macintosh internal clock. In the future, you'll probably want to change the clock time—for example, to adjust for daylight savings time.

Luckily, setting the clock is easy. You'll see how easy now.

Choose Set Time from the Interface menu.

A dialog box appears, asking if you want to change the interface clock to Macintosh time. The dialog box shows you the time according to the interface and the time according to the Macintosh clock.

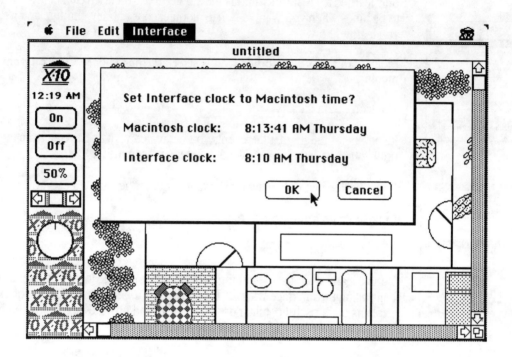

Note that if the Macintosh clock time is not correct, you should *not* set the interface clock until you've changed the Macintosh time in the Control Panel (see your Macintosh owner's manual if you aren't sure how to change the Macintosh time).

If you've completed the preceding sections of this chapter, you've already set the interface clock time, and the two times shown should be about the same.

Click OK.

Changing the Background

Many people open the Home Automation software, look at the background, and immediately say, "My house doesn't look anything like that." Luckily, you can change the background however you want. The background can be any MacPaint document.

Several different backgrounds are included with the Home Automation software. The instructions that follow show how to use one of these backgrounds. But first, you'll need to close the current background.

Choose Close Background from the File menu.

The background disappears, and you are left with only your single module icon.

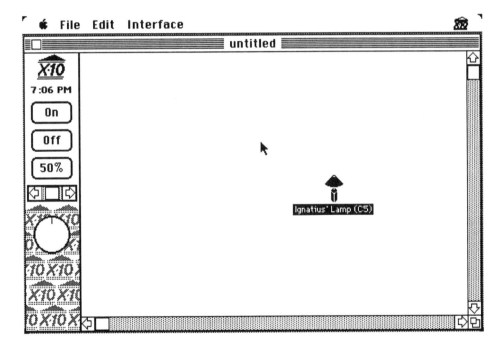

Note that the module icons do not disappear with the background. X-10 (USA), Inc. realized that you might want to change backgrounds without necessarily changing module icon information. Otherwise, you'd have to reenter all of your module icon information every time you changed backgrounds—a very tedious process.

To give you maximum flexibility, the Home Automation software stores information in three places: Backgrounds, the Modules file, and X-10 documents. Each is described below.

Floor Plan

Backgrounds are MacPaint documents that give you a reference point for the Module Map. They can help you visualize your home and the location of modules in it. The backgrounds are separate entities, so you can change them without changing either the module icons or the events you've created.

Modules

The Modules file contains information on the modules you've created and the latest background you've chosen for the X-10 software. The Modules file is created by the Home Automation software *automatically*. There is only one Modules file, which you should leave alone unless you want to erase *all* module information. In that case, throw the Modules file in the Trash.

untitled

X-10 documents are created by the Home Automation software. They contain information about the timed events you have created for the modules in your system.

You create X-10 files with the Home Automation software, and you can have several of them. Here's why you might want more than one. Generally, the modules you install in your home will stay fixed. That is, you install modules for lamps, appliances, sprinklers, and so on, and you don't change them. Events, however, can change for a number of reasons. For example, you might want the lights to go on and off at different times in the winter and the summer. In this case, you could create one X-10 file called Winter and another called Summer. But in both cases, the modules you've installed would stay the same.

You've just closed the background you were using, but the Modules file and the X-10 file haven't changed, so you are working with the same module icons and the same timed events.

Now you're ready to open a different background.

Choose Open Background from the File menu.

You can choose any MacPaint document to be your current background. Now you'll look at the Rooms background.

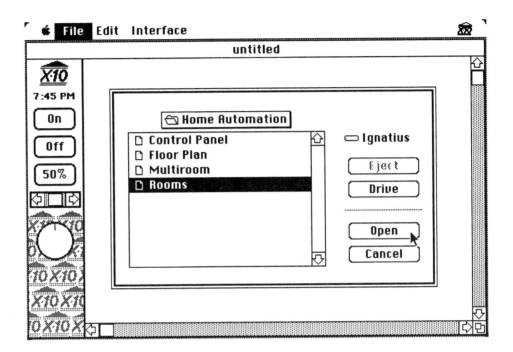

Select Rooms by clicking it. Then click Open.

A new background appears. Notice that your module icon is still in the same spot, and your Event List is the same.

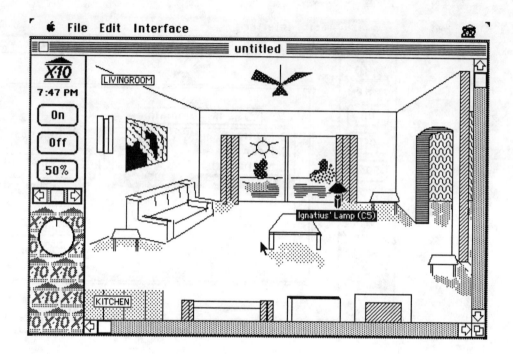

It's important to remember that if you don't like any of the backgrounds that come with the interface, you can easily create your own with a graphics program such as MacPaint. Although the backgrounds supplied with the interface are very attractive, you might be happier with a background that's more like the floor plan of your house. That way, you'll be able to easily identify and work with the icons you create.

Printing

There may be times when you'll want to print a list of the module icons you've created, and the timed events you've set for each one. For example, when you are creating a new X-10 file of timed events, you might want to refer to the current interface settings so you know what to change.

Printing a list with the Home Automation software is very easy.

Choose Print from the File menu.

A dialog box appears in which you can enter printing options.

Click OK.

In a few moments, you'll have a printout of information about the module icon you created. Your printout should look similar to the one shown below.

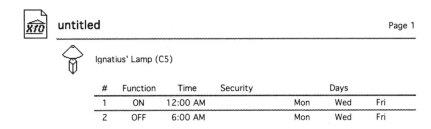

Saving Files

Now you're ready to quit the Home Automation software, but first you'll want to save the events you've created in an X-10 file.

Choose Save from the File menu.

A dialog box appears asking you for a name for the document.

Type a meaningful name for your document, then click Save.

Now you're ready to quit the application.

Choose Quit from the File menu.

You are presented with a dialog box asking if you want to send events to the interface before quitting. This reminder appears because you've modified the Event List of one or more module icons without sending the updates to the interface.

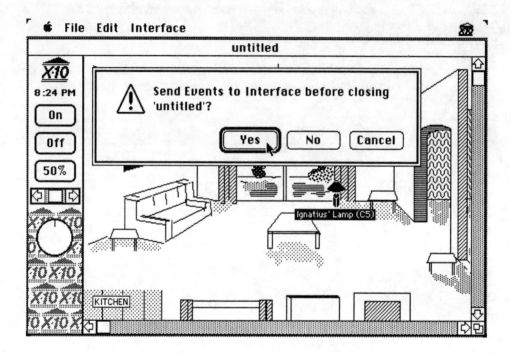

Generally, you'll want to update the interface to include the work you've done. Because this is just an exercise, though, you'll forgo sending events this time.

Click No.

In a few seconds, you'll return to the Macintosh desktop.

Remember that when you changed the Base Housecode of the interface to C, you erased all information in the interface. Now, when you come back to the interface to begin entering real information about your setup, you won't have any "baggage" lying around from the exercise. You'll be starting with a clean slate.

Clearing the Interface

If you ever want to clear the interface completely of all timed events, all you need to do is remove the battery and unplug the interface for a few seconds. This will clear all of the interface memory.

Moving the Interface

Once you've updated the interface with the "Send Events to Interface" command, you can move the interface anywhere and it will still execute your timed events. That is, you don't have to leave it connected to your computer. The only time you need to connect the interface to your Macintosh is when you want to update the timed events in the interface or when you want to control devices immediately.

Instant X-10

From time to time, you might want to turn a lamp or appliance on or off without running the Home Automation software. For example, if you're working at your desk and your spouse asks you to turn on the heater upstairs, it's a little inconvenient to quit what you're doing, run the X-10 software, find the module icon for the heater, select it, and click ON just to turn the heater on. Fortunately, there's a better way.

With the **Instant X-10 Desk Accessory,** you can turn lights or appliances on or off without leaving the application you're using. You just open the desk accessory from within your application, click a few buttons, and you can instantly control lights and appliances in your home.

Installation

Before you can start using Instant X-10, you must install it.

If you are using Macintosh system software version 7.0 or later, install Instant X-10 by dragging its icon from the X-10 disk into your closed System Folder.

The Instant X-10 Desk Accessory is automatically installed in the appropriate location in your System Folder.

If you are using a version of the Macintosh system software earlier than version 7.0, install Instant X-10 using the Font/DA Mover.

If you are not sure how to use the Font/DA Mover, refer to the manuals that came with your Macintosh computer.

Starting Instant X-10

You start Instant X-10 the way you start any Macintosh desk accessory: by choosing it from the Apple () menu.

Choose Instant X-10 from the Apple menu.

The X-10 Desk Accessory leaps onto the screen. The major components are described below.

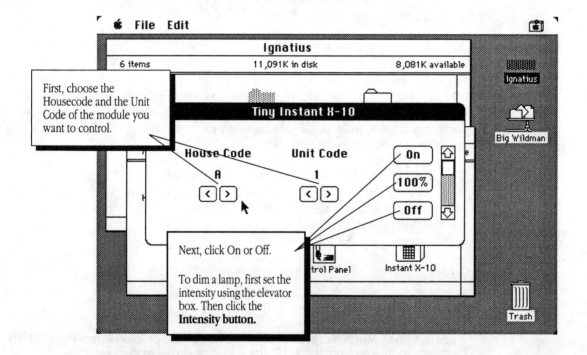

Using Instant X-10 is easy.

Select the Housecode and the Unit Code of the module or modules you want to control.

You select the Housecode and Unit Code using the "<" and ">" buttons, which cycle the Housecode from A to O and the Unit Code from 1 to 16.

Next press On or Off.

To turn the selected lamp or appliance on, just press On or Off. This is not rocket science.

If you want to dim a lamp, you first need to select the intensity you want.

Use the elevator box to set the intensity.

Initially, the elevator box is at the top, and the intensity is 100%. As you drag the elevator box down, the intensity decreases. The current intensity is always shown in the Intensity button.

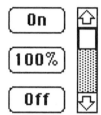
Initially, the elevator box is at the top, and the intensity is 100%.

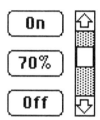
As you drag the elevator box down, the intensity decreases.

Once you've set the intensity, click the Intensity button to dim or brighten the lamp.

You have now learned everything there is to know about Instant X-10.

Summary

You've finished the chapter. Congratulations. If all's gone well, you've learned to

- Set up and test the Home Automation Interface.
- Install the Home Automation software.
- Create module icons, and use them to control modules.
- Create timed events.
- Review timed events stored in the interface.
- Set the Base Housecode so you can use the rocker buttons on the top of the interface.
- Change the floor plan by selecting a new background.
- Print information about the module icons and timed events you've created.
- Use the Instant X-10 Desk Accessory to control modules from within any Macintosh application.

7 IBM PC

Overview

Assuming you haven't skipped right to this chapter, you've already learned a lot about home automation. In fact, you already know enough to build a very functional system that will control lights, appliances, even a security system.

In this chapter, you'll learn how you can control a reasonably complex home-automation system with an X-10 Home Automation Interface and an IBM personal computer or compatible computer.

Using the interface with an IBM PC gives you several advantages over systems that use the other controllers described in this book:

- You can control modules set with any Housecode and any Unit Code; this means that you can control up to 16 x 16 = 256 devices!
- You can create up to 128 timed events.
- You can "program" each module to go on and off at specific times, and on specific days of the week.
- You can program Lamp Modules to dim or brighten to a certain intensity at a given time. This might be done, for example, to add a technical touch to an otherwise very romantic moment.

As you go through this chapter, you'll first learn how to attach the Home Automation Interface to your computer. Then you'll learn how to set up the Home Automation software and use it to enter information about the modules you've installed. Next, you'll learn how to erase module information, and how to control modules either immediately or with timed events. Finally, you'll learn how to save timed events in a file, how to print lists of timed events, and how to exit from the program.

When you're finished reading this chapter, you should be able to use your computer to turn devices on and off both immediately and by using timed events. You'll also know how to save groups of events in a file that you can read into the computer and load into the interface.

This chapter assumes that you are familiar with the operation of your computer. If you aren't, review your owner's guide before continuing the rest of this chapter.

Setup

Before you try to use the interface with your computer, you'll need to ensure that the interface is functioning properly. Here's how to do that:

- Set the address of a Lamp Module to A1, attach the module to a lamp, and then plug the module into an electrical outlet.
- Next, plug the interface into a different electrical outlet. (Although you can use any electrical outlet in the house, choose one in the same room so you don't have to run around.)
- Press the top of the rocker button labeled "1" on the interface to turn the lamp on. Once the lamp has turned on, press the bottom of the same button to turn the lamp off.
- Set the Lamp Module to A2, and repeat the preceding step with rocker button 2.
- Continue as outlined above, testing rocker buttons 3 through 8.

If you experience any problems, be sure you have set the Lamp Module correctly. If you still have problems, call X-10 (USA) Inc.'s Customer Service Department for help (the telephone number is given in Appendix C, page 173).

One last thing you should do before connecting the interface to your computer is to install a 9-volt battery in the battery compartment on the back of the interface. The battery will provide backup power to the interface when it is not plugged into an outlet or when the electricity is off. Without the battery installed, you'll have to reprogram every timed event whenever the interface is unplugged from an electrical outlet or the power goes off.

Connecting the Interface to Your Computer

The interface works with IBM PC (including PS/2) computers and IBM-compatible computers that have a serial (RS-232) card installed in them, or come with a built-in serial port.

Before you do anything else, you'll need to connect the interface to your computer.

Connect the interface to the RS-232 serial port on your computer, using the cable that comes with the interface.

If you are not sure how to do this, consult the manual that came with your computer. If you have an IBM PC XT, PC AT, or IBM compatible, you may need an additional 9-pin to 25-pin converter cable.

Next, you'll copy the MS-DOS system files to the X-10 Home Control disk.

Preparing the Software (No Fixed Disk Drive)

Follow these instructions if your computer does not have a fixed disk drive (also called a hard disk drive). If you do have a fixed disk, skip ahead to the next section.

First you'll need to start the computer.

Insert a DOS system disk in drive A. Then turn on your computer.

Wait for the A> prompt before continuing. Before you can use the interface software, you need to prepare it for use as a system disk.

Insert the X-10 Home Control program disk in drive B. Then type "SYS B:". Finally, press Enter.

Wait again for the A> prompt, which is your signal that the system has finished copying the necessary files to your Home Control disk.

Type "COPY COMMAND.COM B:". Then press Enter.

That's it! Now the disk is ready for use. It's a good idea to make a backup of the prepared disk and store it in a safe place. Once you've done that, you're ready to use the software.

Remove the DOS system disk from drive A. Then remove the X-10 Home Control disk from drive B and place it in drive A.

You do not need to read the next section, which gives installation instructions for users who have a fixed disk drive. Instead, skip to "Starting the Program" on page 136.

Preparing the Software (With Fixed Disk Drive)

The first thing you need to do is create a subdirectory and copy the Home Control software into it.

Turn the computer on. When you see the C> prompt, create a directory named "X10" by typing "MKDIR X10" and pressing Enter.

Next, change to the directory you just created.

Type "CD X10" and press Enter. Next, place the X-10 Home Control disk in drive A and type "COPY A:*.*"

That's it. The Home Control software is on your fixed disk drive. In the future, when you want to use the X-10 software, follow these steps:

- Turn the computer on and wait for the C> prompt.
- Type "CD X10" and press Enter.

Now you're ready to use the software.

Starting the Program

To help you learn about the Home Control software, the following sections will lead you through a series of exercises. You'll use the interface and Home Control software to control a single lamp. To do the exercises, you should get a Lamp Module, set its address to "C5," plug a lamp into the module, and then plug the module into an electrical outlet in the room. Once you've done this, you're ready to proceed.

Of course, the first thing you need to do is start the software.

Type "X10" and press Enter.

The first thing the program shows you is the time stored in the interface. If the interface has been turned off, or if this is the first time you are using it, the time will be wrong.

```
     X-10 INTERFACE
(c) 1986,1987,1988,1989  X-10 (USA) Inc.
The interface contains no data

Enter time(12:00 AM MON) : _
```

Enter the current time and the day of the week.

Be sure to enter the hours, minutes, AM or PM, and the day of the week. The day of the week should be a three-letter abbreviation: SUN, MON, TUE, WED, THU, FRI, or SAT.

Next, the software will report the **Base Housecode.** The Base Housecode is the Housecode that will be used when you press the rocker buttons on the front of the interface to operate lights and appliances. You can use any Housecode with the rocker buttons, but you can choose only one. The default, or preselected, value for the Base

Housecode is "A." At this point, you'll just use the default value. This isn't a final decision: You'll have a chance to change the Base Housecode every time you start the program.

```
        X-10 INTERFACE
 (c) 1986,1987,1988,1989  X-10 (USA) Inc.
 INTERFACE UPLOAD TIME is : 12:13 AM gra

 Enter time(12:00 AM MON) : 7:15 PM SUN

 INTERFACE BASE HOUSECODE is  : A

 Want to change it ? (Y/N)
```

Press "N" to respond no to the question, "Want to change it?"

Entering Module Information

Now you're ready to enter information about the modules you have installed. Then you can control them either immediately or by creating timed events.

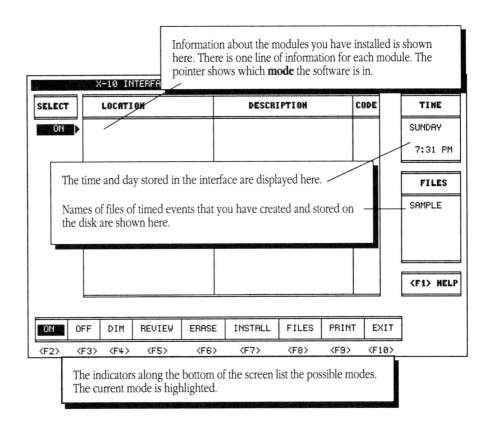

In the center of the screen are three large boxes. They give information about the modules you have installed. You haven't installed any modules yet, so the boxes are empty. To the left of the three boxes is the **pointer.** It shows you which module you are currently working with.

To the right of the three large boxes are two smaller boxes. The first shows the time and day currently stored in the interface. The second shows the files (collections of timed events) that you have created. Because this is the first time you've used the software, only one file is shown—the sample file that ships with the Home Control software.

At the bottom of the screen are ten indicators that show the different software modes you can use. The function key equivalents are shown beneath the modes. Here's a quick description of the modes:

ON, OFF, DIM Lets you send X-10 commands to the module that is selected with the pointer.

REVIEW, ERASE, INSTALL Lets you "install" modules by entering information about them; also lets you erase and change module information and review the list of timed events.

FILES Lets you work with files that contain stored information about modules and timed events.

PRINT Prints a list of all the module information you have created, along with a list of timed events for each module.

EXIT Allows you to exit from the program.

You change the mode by pressing the right (→) and left (←) **arrow keys**.

Press the right arrow key (→) once.

The software changes to Off mode.

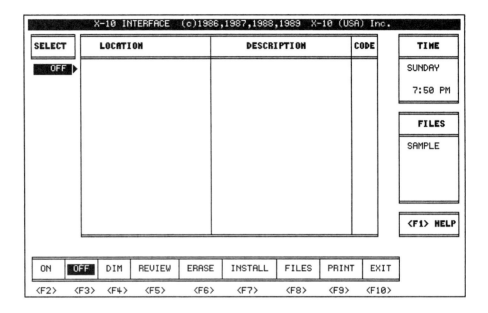

Note that both the indicator at the bottom of the screen and the pointer change to show that the software is in Off mode.

Install mode allows you to enter information about new modules. Because this is the first time you've used the interface, you need to enter information.

Press the right arrow key four times to select Install mode. Then press Enter.

Note that you can also press the function key labeled "F7" to select INSTALL. If you use F7, you don't need to press Enter.

Above the mode boxes you will now see the prompt "Enter the LOCATION:". Here you enter the location of the module you want to control—for example, Kitchen, Living Room, Front Yard, and so on.

Type "RIGHT HERE" and press Enter.

Notice that RIGHT HERE appears in the Location area, and the pointer box moves to the Description area. Now the prompt at the bottom of the screen reads "Enter the DESCRIPTION:".

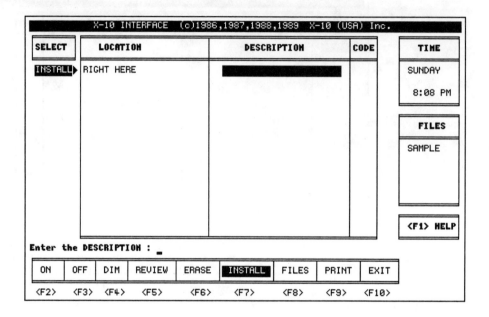

Here you'll type a description of the module you are controlling.

Type a name for the module. Then press Enter.

Your description is entered into the appropriate box, and the prompt now reads "Enter the code (A 1)." Now you must enter the address (Housecode and Unit Code) of the module you are controlling. The prompt will always show you the address of the next available module, in this case A1. If you want to accept that address, just press Enter. Now, however, you want to set the address to C5, so you can control the lamp you set up earlier.

Type "C5". Then press Enter.

Congratulations! If all has gone well, you have just installed your first module. Your screen should look like the one shown below.

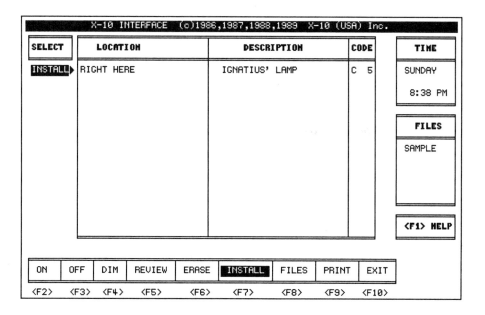

Erasing Module Information

If your screen does not look like the one shown above, then you need to erase the line that you've just created and start over. Here's how to erase a line.

- Use the left and right arrow keys to select Erase mode (to the left of INSTALL), then press Enter.
- You'll be asked if you really want to erase the line. Type "Y," for yes, to confirm that you want to erase it.
- Press the right arrow key once to return to Install mode; then press Enter.
- Follow the instructions above to enter information about the module.

Controlling Modules

Now that you've installed a module, you need to know how to do something with it. The first thing you'll do is turn on your lamp. To do this, you need to be in On mode.

Press the left arrow key until the software is in On mode. Then press Enter.

The mode indicators change to time indicators, allowing you to select the time you want the event to happen. Your options are described below.

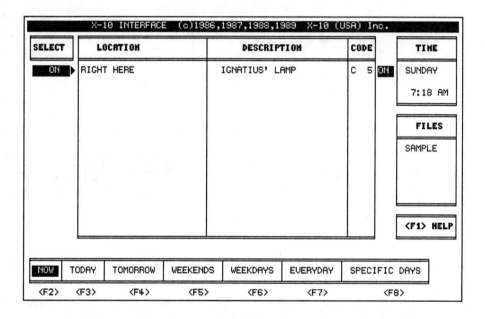

NOW	Turns the module on or off immediately.
TODAY	Prompts you for a time; the event will happen once within the 24-hour period beginning at midnight on the day you create the timed event.
TOMORROW	Prompts you for a time; the event will happen once within the 24-hour period beginning at midnight on the day after you create the timed event.

WEEKENDS Prompts you for a time; the event will happen every Saturday and Sunday of every week until you erase the event.

WEEKDAYS Prompts you for a time; the event will happen every Monday, Tuesday, Wednesday, Thursday, and Friday of every week until you erase the event.

EVERYDAY Prompts you for a time; the event will happen every day of every week until you erase the event.

SPECIFIC DAYS Prompts you for the days of the week on which you want the event to occur. Once you enter the days, you'll be asked for a time; the event will occur on the days you specify, every week until you erase the event.

You'll create a timed event in a little while. First you'll turn the lamp on.

If necessary, use the left and right arrow keys to select NOW. Then press Enter.

A message is displayed while the interface sends an address command followed by an ON function command to the module. Shortly after the message disappears, the lamp should turn on.

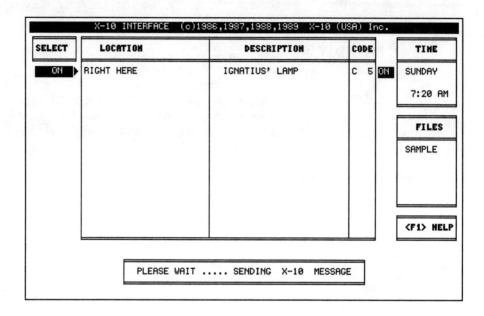

If the lamp doesn't turn on, make sure the Housecode on the module is C and the Unit Code is 5.

Using the Function Keys

By now you've probably decided that pressing the left and right arrow keys four or five times to do things is a little annoying. Luckily, there's an easier way: the function keys.

Look at the bottom of your screen.

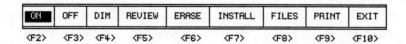

Below each mode indicator is a function key designator—<F2>, <F3>, and so on. Each designator refers to one of the function keys at the top of your keyboard. For example, <F2> represents function key F2, and so on. To change modes, you can either press the left and right arrow keys to select the mode, and then press Enter; or you can press the appropriate function key.

For example, to select DIM, you could either press the right arrow key twice and then press Enter, or you could just press F4.

Press the key labeled "F4" at the top of your keyboard. Do not press Enter.

The mode immediately changes to Dim mode.

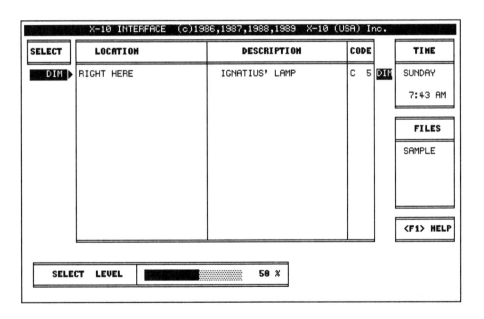

Using the function keys is much faster than selecting modes with the left and right arrow keys. From now on, we'll give instructions using the function keys, but you can use either method.

To dim a lamp, you select the level of intensity by adjusting the sliding scale at the bottom of your screen.

You adjust the scale using the right and left arrow keys. Pressing the left arrow key once reduces the intensity by 10%. Pressing the right arrow key once increases the intensity by 10%.

Press the left and right arrow keys, watching the intensity indicator adjust at the bottom of your screen. When you are finished, change the intensity back to 50%. Then press Enter.

You'll dim the lamp now.

Press F2 to dim the lamp.

The words "Please wait...sending X-10 message" appear on your screen briefly; then the lamp dims to 50% of its original intensity.

Setting Timed Events

Now that you've mastered ON and DIM, you'll use the Off mode to turn off the lamp in your room. This time, however, you'll set a timed event. That is, you'll have the lamp turn off at a specific time in the future.

Press F3 to select Off mode.

You are presented with the same choice of times that you saw for ON and DIM. This time, you'll choose TODAY instead of NOW.

Press F3 to choose TODAY.

A screen appears, asking you to enter the time you want the lamp to turn off.

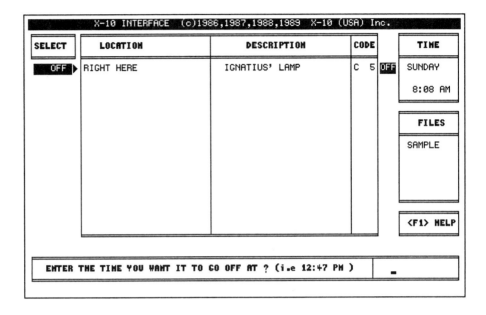

Now look at the Time box in the upper-right corner of your screen. This tells you the time stored in the interface. For the screen above, the time is 8:08 AM. The time on your screen will probably be different (unless you happen to be doing this at 8:08 AM!).

You're going to set the lamp to go off five minutes from now; that is, five minutes past the time shown in the Time box on *your* screen.

Enter a time five minutes past the time shown on your screen. Then press Enter.

Note that you must enter the time and AM or PM. If you do not, the interface software will not accept your entry, and it will prompt you to enter another value. Also, be sure to enter a time five minutes past the time shown on your screen, not the time shown in the screen example above. Otherwise, you might be waiting a very long time for the light to turn off!

Once you have entered a time in the right format, the software accepts it, and asks you if you would like to program an On time.

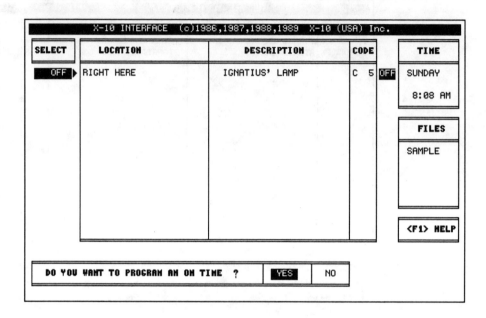

The software designers realize that when you create an event to turn something off, you'll probably also want to turn it on, so they offer you the option to enter On mode right after you program an Off event. Right now, let's just concentrate on getting this first event to work.

Press the right arrow key to select NO. Then press Enter.

You return to the main screen, but you should notice one difference: Now the name of your module is in boldface.

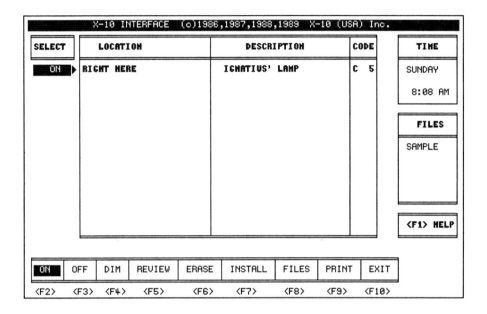

This is a reminder that you have programmed one or more events for that module. This may not seem like a big deal now, because you've installed only a single module. But in the future, after you've installed a screenful of modules, this will be a useful visual clue to show you which ones have been programmed with timed events.

Now you should watch the time on your screen. In five minutes, when the interface clock reaches the time that you set for the lamp to turn off, the lamp should, indeed, turn off. If it doesn't, check all the usual possibilities:

- Make sure that the Housecode and Unit Code on the module are set correctly.
- Make sure that you have correctly set the time for AM or PM.
- Make sure that the interface is still plugged into the wall.

If the lamp still doesn't turn off, there's a chance that you set the time incorrectly. For example, it was 8:08 AM when I set the Off time, so five minutes later would be 8:13 AM. But what if I typed 9:13 AM? Either I'd have to wait an hour for the lamp to go off, or I'd have to change the timed event. You'll learn how to modify existing events now.

Reviewing and Deleting Timed Events

Looking at the mode indicators at the bottom of your screen, you've probably already figured out that you'll use Review mode to review timed events.

Press F5 to enter Review mode.

You are presented with another screen that allows you to review the timed events you've set for the selected module. At the top of the screen, in boldface, are the location, description, and address of the module you are reviewing. Below that are four columns that contain the function, time, mode, and "when" for each of the timed events that you've set. There is one line for each timed event. Finally, at the bottom of the screen, there are three options for you to choose from.

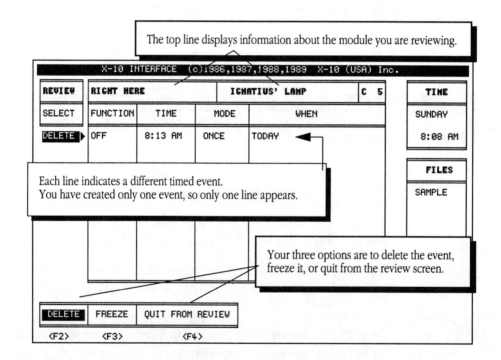

DELETE Allows you to delete the timed event; the function key equivalent is F2.

FREEZE Stops an event from occurring until you unfreeze it. You unfreeze an event by selecting it with the up or down arrow keys and then selecting FREEZE again. The function key equivalent is F3.

QUIT FROM REVIEW Returns you to the main screen; the function key equivalent is F4.

Tip: When you are in Review mode, you can review timed events for other modules without first returning to the main screen. To do this, hold down the Shift key and press the up or down arrow keys to scroll through the list of events for all of your modules.

If you make a mistake, press F2 to delete the timed event. Then follow the instructions in the previous section to re-create the timed event. If you have not made a mistake, and everything is OK, press F4 to return to the main screen.

Saving, Printing, Exiting, and Other Activities

Saving Timed Events in a File

Now you are almost ready to enter information for all the modules you've installed in your home and to create timed events for everything in sight. Already, you can probably think of occasions when you'll want to change your timed events. For example, during the summer months, you'll want the lights to come on later in the evening, and if you're controlling a heater, you probably won't want it to come on at all. Likewise, in the winter, you'll want to turn the lights on earlier in the evening, have the heater on most of the time, and leave the air conditioner off.

One solution is to change the timed events you've programmed each time the seasons change. This is somewhat inefficient, though, because you end up retyping things over and over again. Fortunately, there's a better way. You can save a list of timed events to use at a later time. For example, you could save one set of events called "SUMMER" and another called "WINTER" that you swap from season to season.

To create a file, you first select the Files mode.

Press F8 to select Files mode.

A new screen appears.

```
┌──────────────────────────────────────────────────────────────────────┐
│         X-10 INTERFACE  (c)1986,1987,1988,1989  X-10 (USA) Inc.      │
├────────┬──────────────────┬──────────────────────┬──────┬────────────┤
│ SELECT │     LOCATION     │     DESCRIPTION      │ CODE │    TIME    │
│        ├──────────────────┼──────────────────────┼──────┤            │
│        │ RIGHT HERE       │ IGNATIUS' LAMP       │ C  5 │   MONDAY   │
│        │                  │                      │      │            │
│        │                  │                      │      │   8:00 PM  │
│        │                  │                      │      ├────────────┤
│        │                  │                      │      │   FILES    │
│        │                  │                      │      │            │
│        │                  │                      │      │   SAMPLE   │
│        │                  │                      │      │            │
│        │                  │                      │      │            │
│        │                  │                      │      │            │
│        │                  │                      │      ├────────────┤
│        │                  │                      │      │ <F1> HELP  │
├────────┼──────┬───────┬───┴──┐                   │      │            │
│  LOAD  │ SAVE │DELETE │ QUIT │                                        │
│  <F2>  │ <F3> │ <F4>  │ <F5> │                                        │
└────────┴──────┴───────┴──────┘
```

You have the following options:

LOAD Load a file from disk. You'll be asked to enter the filename of the file you want to load. Type the name and press Enter. Note that when you load a file, the contents of the file completely replace everything that is stored in the interface. So, for example, if you've just created a long list of timed events, you won't want to load a new file until you save the current events in a file, or you'll have to retype them. The function key equivalent is F2.

SAVE Save the current set of timed events. You will be asked for a filename. Type the name and press Enter. For example, you might type FIRST as a filename for the timed event you just created. Note that your files will be saved with the extension ".X10." For example, if you save a file and type "FIRST" for the name, the complete file name will be "FIRST.X10." The function key equivalent is F3.

DELETE   Delete a file. You will be asked for the name of the file to delete. Type the name and press Enter. The function key equivalent is F4.

QUIT   Return to the main program screen. The function key equivalent is F5.

Now let's return to the main menu.

Press F5.

Printing Files

You can print the current file by choosing PRINT. This gives you a printed list of all the modules you've defined, and all the timed events you've created for each module. A printed list can be very useful when you're creating files for different seasons of the year, or if you want to quickly review all the timed events you've programmed before going on vacation, to see if you should make any changes before you leave.

Press F9 to print the current file.

Your printout should look like the one shown below.

```
   X-10 INTERFACE
 ---------------------------------------------------
 ---------------------------------------------------
 RIGHT HERE            IGNATIUS' LAMP          C  5
    OFF  8:13 AM  ONCE      TODAY
 ---------------------------------------------------
```

Exiting from the Program

You exit from the program by choosing EXIT from the main screen or by pressing F10.

Press F10 to exit from the program.

You're back at the familiar DOS C> prompt (or the A> prompt, if you don't have a fixed disk).

Clearing the Interface

If you ever want to completely clear the interface of all timed events, all you need to do is remove the battery and unplug the interface for a few seconds. This clears all of the interface memory.

Moving the Interface

Once you've opened a file, the timed events contained in the file are automatically loaded into the interface. Then you can move the interface anywhere and it will still execute your timed events. That is, you don't have to leave it connected to your computer. The only time that you need to connect the interface to your PC is when you want to update the timed events in the interface or when you want to control devices immediately.

Where to Go from Here

Now you should know everything you need to build a complete home-automation system around the X-10 Home Automation Interface. As you add module information to the system and create timed events, you should be able to find all the information you need in the sections that you've just read.

As always, if you have problems, see Appendix B, "Troubleshooting," on page 163.

A Technical Overview

Overview

You don't need to refer to this appendix when designing, installing, and using X-10 systems. It's for the curious, who can't merely plug something in without wondering how it works (like us). If you're an electrical engineer, you'll be severely disappointed by the depth of the information, but for weekend soldering-iron jockeys, the appendix should provide a general understanding of how X-10 works.

How It Works

X-10 is based on a technology broadly called **Power Line Carrier.** As you have seen, this design is based on the concept of using the existing transport mechanism for electricity, such as the AC power lines in your home, to carry the commands that controllers send and modules receive. The most obvious benefit of this technology is that there's no need for additional wiring, which would be a huge obstacle in automating existing homes. Power Line Carrier systems also draw the power necessary for the signal—a tiny amount—directly from the power line that is used as the transport mechanism. No alternate power or transformer is required.

All this takes place with no noticeable effect on the power line itself. The signals are sent back and forth at a frequency much higher than the frequency of the electrical current. Appliances and lights that are not connected to X-10 modules will never know the difference. The exceptions are other devices that use the same system to transmit signals, such as certain home intercom systems. Problems can occur if the intercom uses frequencies in the same bandwidth as X-10. Most manufacturers attempt to avoid this problem, because the X-10 standard is so well established.

Household wiring is subject to "noise" on the line from appliances, fluorescent lights, electric motors, televisions, and other items that create electrical noise while operating. This limits the amount of information that can effectively be transmitted in this manner.

For systems like X-10, the amount of data sent is actually quite small. The signal is sent in 1-millisecond bursts of 120 kHz. The information is sent at the **zero crossing point** of the 60-Hz frequency of electrical power (see the figure on the next page). A 120-kHz signal at the zero crossing point represents a **binary** "1," and the lack of a signal at the zero crossing point represents a binary "0."

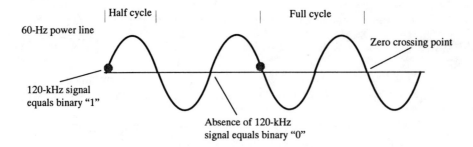

You might think of this as an inaudible Morse code signal traveling across your electrical wires.

When a module "hears" the **Start Code**—1110—come across the line, it knows that the next signal will be the Housecode. If the next thing the module hears is its own preset Housecode, it listens to the next four signals, or bits, to find out if its unique Unit Code is being sent. If it hears that code (represented by Unit Code 1 through 16 on each module), it is ready to act on the next code sent, which is the five-bit Function Code. If another Housecode or Unit Code comes across the line at any point in the sequence, the module recognizes that the message is meant for another module, and stops.

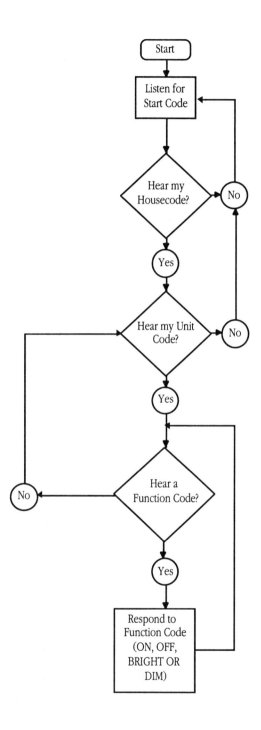

The complete transmission of the X-10 code requires 11 cycles of the power line. The first two cycles represent the Start Code, the next four cycles represent the Housecode, and the last five represent either a Unit Code (1 through 16) or a Function Code (ON, OFF, DIM, and so on). Modules listen to Function Codes only if they have received a correct Housecode and Unit Code. The order is critical—a complete code block with the Unit Code first, followed by a complete code block including the Function Code.

Each complete code block is transmitted twice, with three clear (no signal) power-line cycles between pairs. BRIGHT and DIM Function Codes are sent continuously, with no cycles separating them, since they are, in effect, "increasing" or "decreasing" commands instead of single ON or OFF commands. For example, if BRIGHT is sent and the next code is DIM or Unit Code 1, then a three-power-line-cycle gap will be left.

Each bit or signal is actually transmitted twice—once in its true form and once in its complement, or opposite, form on the next half of the cycle of the power line. This helps ensure the accuracy of the signals. If any signal on the first half of the power cycle is not followed by its complement on the next half of the power signal, it's ignored. This is a very simple method of error checking that helps ensure that your modules don't respond to line noise.

The entire code sequence, including complements, is as follows:

Start Code	H1 $\overline{H1}$	H2 $\overline{H2}$	H4 $\overline{H4}$	H8 $\overline{H8}$	D1 $\overline{D1}$	D2 $\overline{D2}$	D4 $\overline{D4}$	D8 $\overline{D8}$	D16 $\overline{D16}$
1110	1 0	0 1	1 0	1 0	0 1	0 1	0 1	1 0	1 0
1110	1	0	1	1	0	0	0	1	1

Each signal, such as H1, is followed by its complement, $\overline{H1}$. If H1=1—meaning that a 120-kHz signal was detected at the crossing point of the 60-Hz power line, and on the next half cycle another 120-kHz signal was detected—a transmission error (no complement) would be detected, and the entire code sequence would be ignored. The exception to this rule is the Start Code, which is always the unique 1110. Since no complement is necessary for the Start Code, four bits of information can be sent in two cycles of the power line (one bit per half cycle, with no complement). The next bit requires a full cycle for the signal on the first half and the complement on the second half.

Each complete 11-cycle code block (Start Code, Housecode, and Function or Address Code) is sent twice to double the chances that the code is received correctly. The two types of redundancy (sending the complement bit and sending the complete code block twice) are a simple yet effective way of ensuring signal accuracy and transmission without requiring a "return signal" from the modules, which would require additional hardware and dramatically increase cost.

The code sequence for address commands is shown below.

	Housecodes					Unit Codes				
	H1	H2	H4	H8		D1	D2	D4	D8	D16
A	0	1	1	0	1	0	1	1	0	0
B	1	1	1	0	2	1	1	1	0	0
C	0	0	1	0	3	0	0	1	0	0
D	1	0	1	0	4	1	0	1	0	0
E	0	0	0	1	5	0	0	0	1	0
F	1	0	0	1	6	1	0	0	1	0
G	0	1	0	1	7	0	1	0	1	0
H	1	1	0	1	8	1	1	0	1	0
I	0	1	1	1	9	0	1	1	1	0
J	1	1	1	1	10	1	1	1	1	0
K	0	0	1	1	11	0	0	1	1	0
L	1	0	1	1	12	1	0	1	1	0
M	0	0	0	0	13	0	0	0	0	0
N	1	0	0	0	14	1	0	0	0	0
O	0	1	0	0	15	0	1	0	0	0
P	1	1	0	0	16	1	1	0	0	0

And the code sequences for Function Commands are shown in the table below.

Function Command	Code Sequence				
ALL UNITS OFF	0	0	0	0	1
ALL LIGHTS ON	0	0	0	1	1
ON	0	0	1	1	1
OFF	0	0	1	1	1
DIM	0	1	0	0	1
BRIGHT	0	1	0	1	1
ALL LIGHTS OFF	0	1	1	0	1

The X-10 code format is patented, but there are ways to use the technology for your own product ideas without dealing with complicated licensing issues. If you're interested in developing your own X-10–compatible products, we suggest that you investigate two products available from X-10 (USA), Inc. for OEM manufacturers and hobbyists: the PL513 transmitter and the TW523 transmitter/receiver. Both products plug into regular AC outlets and use a standard telephone RJ-11 interface to connect to your product or experiment. This method eliminates the need to connect directly to 110 AC current and makes implementation safer and easier.

If you want more information about the technical aspects of X-10 technology and more sophisticated aspects of home automation, you can refer to *Circuit Cellar* magazine or call X-10 (USA).

B Troubleshooting

Overview

In general, X-10 Home Automation products are extremely easy to install and use. Like any electronic equipment, however, they require proper setup and configuration to function properly. Occasionally, you may have a problem with an X-10 system. When you do, this appendix is the first place you should look for a solution.

If you can't find the answer here, you should consider consulting a **gearhead** at your local electronics store. These people are generally more than happy to help you solve your problem.

We've divided all the problems we think you'll ever experience into two categories: things don't work at all, and things don't work correctly. In the following pages, we provide troubleshooting tips for both types of problems.

Things Don't Work at All

If you can't get anything to work, that is, you can't turn modules on or off with your controller, then try these solutions.

Is the controller plugged in?

Shame on you if this is the problem. Please consult pages 10, 14, 19, and 31.

Is the outlet a working outlet (not controlled by a wall switch)?

If the electrical outlet is controlled by a wall switch, and the wall switch is off, the controller won't be able to send commands and the system will not work properly. Make sure you've plugged the controller into a functioning electrical outlet.

Is the Housecode the same on the controller and on the Module that you're trying to control?

The Housecode on the controller must match the Housecode on the module, or the module will not respond to the function commands that the controller sends. Check again, even if you think you've set the Housecode correctly—sometimes the dials don't line up perfectly.

Is the appliance or light you are trying to control turned on?

After you plug the appliance into the X-10 module, make sure the appliance is turned on. The X-10 module works by regulating the amount of power that the device receives. That is, the X-10 module can prevent a device that's switched on from turning on, but it can't turn on a device that's switched off.

If you are controlling a light, does the light work? Does it have a working light bulb?

Make sure that the light functions without the X-10 module attached. If it doesn't, you've found your culprit.

Are you sure you're using the right Unit Code for the module?

If you don't press the correct rocker button or Unit Code button on the controller, the light or appliance will not turn on or off.

If you are using a Mini Controller, a Mini Timer, or a Remote Control, also be sure that the SELECTOR switch is set correctly for the module you are trying to control.

If you are controlling a Two-Way Wall Switch Module or a Three-Way Wall Switch Module, is the slide switch in the On position?

If the slide switch on the bottom of the module is to the left, the module cannot be controlled. Make sure the slide switch is in the right position before you try to control the module.

Does the module work?

X-10 products are well built and reliable. But if you suspect that a module may be defective, swap it with another module, preferably one that you know works properly. If this solves the problem, contact X-10 (USA), Inc. about their in-warranty replacement policy. (The telephone number is given on page 172.)

Does the controller work?

If you have two or more controllers, first try swapping the one you are using for another. Although it's unlikely, the controller may not be functioning properly.

If you are using a Mini Timer, be sure that the MODE switch is set to RUN, or you won't be able to control anything.

Try moving the controller and the module to the same electrical outlet (if you can). It's possible that the X-10 commands are not reaching the module because they are traveling over too long a distance. Note that even if you have a small house, or you've plugged the module and controller into outlets that are reasonably close together, you could still run into a distance problem. Here's why.

When you send an X-10 command by pressing a button on the controller, the command must travel from the outlet the controller is plugged into to the outlet that the module is plugged into. In a standard two-phase wiring system (used in most homes), that signal might have to travel the entire length of the house back to the **circuit breaker.** From there, it would need to run out to the local power source, back along the trunk to the circuit breaker, and then through some of your home's wiring before reaching the module. This distance might be more than a quarter mile! Moving the controller and module to the same outlet will help you rule out this problem. In addition, some companies sell "signal bridges," which could help you solve these problems if you have them. For information on vendors of X-10 compatible products, see Appendix C

Things Don't Work Right

Check this section when things aren't working as well as they should be. For example, if devices turn on or off when they shouldn't.

Could someone else (a neighbor or family member, for example) be controlling the module with an X-10 controller?

That light that seems to come on by itself might be turned on by someone who has discovered an X-10 controller.

Could a Mini Timer or Home Automation Interface be sending the signals?

Did you set up automatic commands to be sent from one of these controllers? They'll continue to send those signals until you program them to stop.

Does your neighbor have an X-10 system?

X-10 controllers send signals over the power line, and those signals continue to travel until they reach a pole transformer or lose strength. Four or five homes may be supplied through the same pole transformer. If you suspect that this is the problem, try switching Housecodes until the problem stops. Or better yet, talk to your neighbors about their system and decide which Housecodes each of you will use.

Do you have a wireless intercom system or baby monitor?

Some intercoms also use the power line to transmit signals. These devices "broadcast" their signal continuously in order to transmit sounds, and can wreak havoc on the X-10 signals that are trying to travel on the same wires. Changing Housecodes may help, but it's best to turn off the intercoms except when necessary, or the erratic behavior will continue.

C Compatible Products

You've learned about the basic X-10 controllers and modules that you can use to build a home automation system. These products are just the beginning. There are literally hundreds of companies that manufacture X-10 compatible equipment, from voice-command controllers to designer-quality wall switch modules. This appendix lists some of the most popular vendors of X-10 – compatible equipment and briefly describes their products. You should find this information helpful in expanding your system beyond the limits of the X-10 products described in this book. Good luck!

Advanced Control Technologies, Inc.

1936-B South Lynhurst Drive
Indianapolis, IN 46241
Telephone: (317) 248-2640
Fax: (317) 248-6898

Product: Powerline Control Components

Advanced Control Technologies sells a line of X-10 modules and controllers for commercial and sophisticated industrial applications. The Powerline Control product line includes commercial and industrial quality X-10 compatible modules, controllers, interface transmitters and programmable controllers. For systems with intermittent problems or long wire runs, Advanced Control Technologies produces couplers and repeaters to extend X-10 signals and to cross phases. They also produce signal meters to test for X-10 commands on the power line, and filters to help eliminate noise.

Apex Security Alarm Products (ASAP)

3301 Bramer Drive
Raleigh, NC 27604
Telephone: (919) 876-0010
            (800) 272-7937
Fax: (919) 850-0977

Product: Apex Security System

ASAP produces a series of sophisticated, multizone security systems that use both hard-wired and wireless connections to protect up to 32 zones. The systems have built-in telephone-access control and can control up to 16 X-10 – compatible modules.

Blue Earth Research

310 Belle Avenue
Mankato, MN 56001
Telephone: (507) 387-4001
Fax: (507) 387-4008

Product: Micro-440

Blue Earth Research's Micro-440 programmable controller allows computer programmers to develop products that have monitoring and control capabilities. The controller and the included software require little or no experience with electronic hardware. An MS-DOS PC or a terminal can be programmed in BASIC and/or assembly language to create a sophisticated data collection and monitoring system. This system can then be used to send X-10 – compatible commands to control lights and appliances based on the acquired information.

Enerlogic Systems, Inc.

P.O. Box 3743
Nashua, NH 03061
Telephone: (603) 880-4066
Fax: (603) 880-8297

Product: Enerlogic ES-1400e Intelligent Controller

The ES-1400e Intelligent Controller gives the user sophisticated yet easy-to-use control over an X-10 home-control system via IBM PC software. The software included with the system uses English words instead of complex numbers and letters to name the modules and events that describe a home-automation schedule. It can also monitor X-10 transmissions across the power line and take actions based on those transmissions, as well as sending timed events. Once the system has been set up, the computer can be disconnected and used for other applications.

G.E. Marshall Marketing, Inc.

6 Corbridge Court
Weston, Ontario, Canada M9R 2N9
Telephone: (416) 244-1075
Fax: (416) 240-1048

Product: Tel-a-Control

Tel-a-Control is an X-10 - based heating-control system. It can detect freezing temperatures in an unoccupied residence and turn on the heat to protect against water-pipe damage. It can also work in conjunction with a Telephone Transponder to allow dial-up control of heating and lighting systems, to prepare your home or cabin for your arrival.

Group Three Technologies, Inc.

2125B Madera Road
Simi Valley, CA 93065
Telephone: (805) 582-4410
Fax: (805) 582-4412

Product: Samantha Home Automation System

Samantha is a combination home communications, home control, and security system. It can control up to 128 X-10 modules. It uses voice feedback to help you control the system, and lets you add customized messages in your own voice. Samantha also integrates a telephone management system that includes a memory dialer, speaker phone, and answering machine with beeperless remote operation. It uses existing telephones and wiring to create an in-home intercom system.

Home Automation, Inc.

2313 Metairie Road
P.O. Box 9310
Metairie, LA 70055-9310
Telephone: (504) 833-7256
Fax: (504) 833-7258

Product: Model 1503 Version 2 Home Automation System

Home Automation, Inc. sells a home-automation interface that manages a maximum of 80 zones and can control up to 128 X-10 modules. It can be controlled externally or from inside your house by any Touchtone telephone. The Model 1503 has built-in voice synthesis to prompt you through its commands. Sophisticated series of X-10 commands can be programmed to be executed from the built-in keypad, from a touchtone telephone command, after a change in security system status, or at a designated time. The Interface also has a built-in modem and software that allows complete control and programming from an IBM PC compatible computer.

Home Automation Systems

21 Seascape Drive
Newport Beach, CA 92663
Telephone: (800) SMART-HM
 (714) 642-6610
Fax: (714) 642-7190

Product: PLATO Intelligent Home Controller

PLATO is a sophisticated, integrated home control system that can both send and receive all 256 X-10 control codes. It has an optional integrated security system, support for multiple heating and cooling zones, multiple-zone audio/visual system control, and a built-in telephone voice mailbox and message forwarding system. PLATO can be controlled by any of the following devices: wireless RF or infrared remote control, wall-mounted and tabletop keypads, telephone, voice, and time or sensor inputs—for example, motion detectors and light, temperature, or moisture sensors. PLATO has an integrated natural voice menu and help system that guides you through operations and can give you voice feedback on the status of virtually any component of the system. The system comes custom-programmed for the end user. PLATO is easy to install and use, requiring the user only to plug PLATO into the power supply and to connect it to a telephone line. Optional equipment includes a software program for the IBM PC that lets the user develop his own system with PLATO functionality.

JDS Technologies

16750 West Bernardo Drive
San Diego, CA 92127
Telephone: (619) 487-8787
Fax: (619) 451-2799

Product: TeleCommand System 100

The TeleCommand System 100 expands the concept of telephone-controlled home automation. Users can control up to 100 electrical devices from any Touch-tone phone, including in-home cordless and external cellular phones. The system plugs into an electrical outlet and a modular telephone jack. It can perform multiple X-10 commands at the touch of a single telephone key. It comes pre-programmed and ready to use, but can also be fully customized by the user. The TeleCommand System 100 has a security system that prevents access by outsiders, and works in conjunction with most answering machines.

Mastervoice, Inc.

10523 Humbolt Street
Los Alamitos, CA 90720
Telephone: (310) 594-6581
Fax: (310) 493-4982

Product: Butler-in-a-Box and Series II

The Butler-in-a-Box and the new Series II from Mastervoice provide complete home control and security via voice commands. The products can recognize four separate voices in any language and respond in a digitized male or female voice. You can use voice commands to control X-10 modules, dial and answer your phone, monitor security, change channels on your TV, and manage the temperature. The Series II also can execute multiple X-10 function commands based on a single spoken command.

Solus Systems, Inc.

4000 Kruse Way Place, 3-160
Lake Oswego, OR 97035
Telephone: (503) 635-3966
          (800) 247-5712
Fax: (503) 635-3004

Product: Solus Personal Control Computer

Using the IBM PC software that comes with it, you can configure the Solus Personal Control Computer to control X-10 modules based on a wide variety of input sensors, including rain gauges, thermometers, barometric- pressure sensors, humidity sensors, and moisture sensors. The software gives a graphical, real-time representation of sensor information and resulting X-10 commands.

Unity Systems Inc.

2606 Spring Street
Redwood City, CA 94063-2430
Telephone: (415) 369-3233
Fax: (415) 369-3142

Product: Unity Systems MCPU2 Control Unit and CRT3 Touchscreen

Unity Systems produces a sophisticated alarm and control system that includes a touch-screen option for easy operation. Besides controlling up to 72 X-10 devices, the Unity System can control heating and cooling in individual rooms or zones, provide complete fire and burglar protection, and control lights and appliances.The system can be run by touch-screen control, a Touchtone telephone (local or remote), or an IBM PC.

X-10 (USA) Inc.

185A LeGrand Avenue
Northvale, NJ 07647
Telephone: (201) 784-9700
Fax: (201) 784-9464

Product: X-10 Product Line

X-10 (USA) is the creator of all the products covered in this book and the keeper of the X-10 technology. X-10 (USA) also offers other products, and is continually working to bring new ones to market. The X-10 product line is the place to start for the basic components of home automation.

Glossary

address
The combination of a module's Housecode and Unit Code—for example, A1.

address command
A command that a controller sends to identify modules before sending function commands. The address command consists of a Housecode and a Unit Code. After receiving the address command, modules with that Housecode and Unit Code listen for subsequent function commands, which they execute.

alarm
A device that creates a sound or other signal based on the status of a receiver.

arrow keys
Keys on the keyboard of IBM personal computers that allow you to navigate through the Home Automation software for the IBM PC.

Base Housecode
The default Housecode for the Home Automation Interface. It determines which modules are controlled by the rocker buttons on the Interface.

binary
A numbering system made up of only 1's and 0's that is used by computer systems and electrical devices.

circuit breaker
A safety switch in your wiring that will turn the electricity off to a certain portion of your home if a problem occurs that increases the amount of current beyond its rated capacity.

command
A message that a controller sends to a module to control it. See also **address command** and **function command.**

Control Area
In the Home Automation software for the Macintosh computer, the area on the screen that includes the interface clock, the Events Meter, and the ON, OFF, and DIM buttons. The other portion of the screen is called the Module Map.

controller
A device that sends an address command and a function command that tell a module what function to perform. Many types of controllers are available, including the Maxi Controller, the Mini Controller, the programmable Mini Timer, and the Home Automation Interface, that works in conjunction with a computer. They all have special features that provide a variety of ways to send address commands and function commands to modules.

decibel (dB)
A measurement of sound. 85 dB is loud; 120 dB is painfully loud.

desk accessory
Software for the Macintosh that can be used while other software is running.

device
A generic term for an entity controlled by an X-10 module—for example, a lamp or an appliance.

Diet Coke
A source of nutrition and caffeine for aspiring authors.

Dimmer Control
In the Home Automation software for Macintosh, a control in the Edit Module Program dialog box that allows you to create a timed event that will dim a light.

electric valve
A device that opens and closes when electrical current is passed through it. Used to start and stop the flow of water.

Events Meter
In the Home Automation software for Macintosh, an icon, or picture, that represents the number of events that have been defined relative to the total number available.

fluorescent
A bulb that creates light by exciting a gas contained in a pressurized tube. Fluorescent bulbs are usually long and thin and produce very little heat. They can be turned on and off by an Appliance Module, but should never be used with a Lamp Module. Fluorescent lights cannot be dimmed.

function command
A command that a controller sends to a module after it has identified the module with an address command. Modules respond to all function commands until they receive another, different address command.

gearhead
A person who is very knowledgeable about things mechanical or electronic. Also known as a nerd, and, less often, a techie.

Housecode
A letter from A through O that is used to identify groups of modules. The Housecode and the Unit Code of a module together make up the module's address. Housecodes provide a useful way of grouping modules; for example, all modules with the same Housecode respond to the ALL LIGHTS ON and ALL UNITS OFF commands.

icon
A small picture that represents an item in the Macintosh environment, such as a hard disk, a folder, a module, and so on.

incandescent
A bulb that creates light by forcing electricity through a filament. Incandescent bulbs produce heat. They can be controlled by a Lamp Module, and can be turned on and off and dimmed.

Install mode
In the Home Automation software for the IBM PC and compatibles, the mode used to enter module information.

Instant X-10
A desk accessory for the Macintosh that's used to control X-10 modules via the Home Automation Interface.

Intensity button
A button that sends a DIM or BRIGHT command to a Lamp Module or Wall Switch Module. The intensity button is labeled with a percentage that represents the intensity of the lamp.

Jones, Chuck
The greatest director of animation ever. Creator of the Road Runner, Wile E. Coyote, Pepe LePew, Marvin the Martian, the best Bugs Bunny cartoons, and many, many others. What does he have to do with home automation? Probably nothing, but he's our hero and it's our book.

low voltage
Less than 30 volts.

magnetic contact switch
A device that uses a magnet to open and close a circuit.

mode
One of a number of states of a software program. The mode determines what actions you can take at a particular time. For example, the home automation software has a mode for entering information and a mode for editing information.

module
An interface box that you connect to lamps, appliances, or other devices to control them. Every module has an address, not necessarily unique, that consists of a Housecode and a Unit Code. Controllers communicate with modules by sending addresses and commands to them.

module icon
In the Home Automation software for Macintosh, an icon that represents a module. You control the module by selecting the module icon and choosing a command to send to it.

Module Map
In the Home Automation software for Macintosh, the area on the screen where you create, manipulate, and view module icons.

momentary switch
A switch that closes an electrical circuit when activated and opens the circuit when released.

normally closed
A switch that is typically connected and the attached circuit is complete.

normally open
A switch that is typically disconnected and the attached circuit is not complete.

open-loop system
A design that relies on one-way communication for signaling. In an open-loop system, no confirmation or return signal is used to acknowledge that a signal has been received. X-10 is an open-loop system.

 pointer
In the Home Automation software for the IBM PC, an on-screen indicator that identifies which module you are controlling and specifies which mode the software is in.

Power Line Carrier
A system that uses existing AC power lines for the transmission of control signals from controllers to remote modules. Power Line Carrier systems superimpose signals in a frequency that doesn't interfere with the transmission of power, and they don't require additional wiring.

radio frequency
A transmission method that uses high-frequency waves for signaling. Radio frequency signals can penetrate walls and glass (depending on power levels) and don't require the receiver to be within the line of sight.

receiver
A device that monitors a sensor and can trigger actions, such as sounding an alarm or turning on a light.

rocker button
One of a group of buttons on some X-10 controllers that are used to send an address command and a function command simultaneously.

sensor
A device that detects a change from the standard state. For instance, a magnetic contact switch is a type of sensor that can detect the opening of a door or window.

sounder
A device that creates a sound when electricity is passed through it.

Start Code
The special signal that lets a module know that the next signal coming is the Housecode. The X-10 Start Code is 1110.

techno-speak
A dialect of English most often spoken by gear heads.

timed event
A command that is stored in the controller and sent to modules at a specified time. Timed events can happen just once or can recur on certain days of the week.

Unit Code
A number from 1 through 16 that is used to differentiate among modules. The Housecode and the Unit Code of a module together make up the module's address. Housecodes provide a useful way of grouping modules because all modules with the same Housecode respond to the same ALL LIGHTS ON and ALL UNITS OFF commands.

Unit Code button
One of a group of buttons on some X-10 controllers. The Unit Code button is used to send an address command.

watt
A measurement of the quantity of electrical current that flows through a circuit.

X-10 Home Automation
Home-automation technology that uses controllers and modules that communicate over existing electrical wiring. X-10 technology is inexpensive, flexible, and easy to use.

zero crossing point
The midpoint of the amplitude of an AC sine wave. This is where X-10 signals are sent.

Index

A

air conditioning, controlling 49, 52–55

Appliance Modules 48–50

appliances, controlling 47–65
    overview 47

B

burglar alarm systems *see* home security

C

commands
    ALL LIGHTS ON 4, 6, 12–13
    ALL UNITS OFF 4, 6, 12–13
    BRIGHT 4
    DIM 4
    OFF 4
    ON 4

compatibility 7

controllers 2–3, 9–36
    overview 9

G

garage-door opener, controlling 61–65

gearhead 163

H

heating, controlling 52–55

Heavy-Duty Appliance Module 55–57

Home Automation Interface 27–28
    comparison with other controllers 28
    for IBM PC *see* IBM PC
    for Macintosh *see* Macintosh
    overview 27–28
    using for manual control 27
    using for timed events 27–28

home security 67–89
    basics 68–70
    burglar alarm interface 70–73
    Door/Window Sensors 81–84
    Miniature Remote Control 80
    options 85–89
        Powerhorn 85–86
        Wireless Motion Detector 86–89
    overview 67
    Remote Control 78–79
    Supervised Home Security System 74–84
        Base Receiver 76–77
    using the system 89–92

Housecodes 5–6

I

IBM PC 133–156
    arrow keys 140
    connecting the interface 134–135
    controlling modules 144–148
    entering module information 139–143
    erasing module information 143
    exiting from the program 155
    function keys 146
    overview 133
    preparing the software 135–136
    printing 155
    reviewing and deleting timed events 152–153
    saving files 153–155
    setting up the interface 134
    setting up timed events 148–151
    starting the program 136–138

Instant X-10 129–131
    installing 129–130
    using 130–131

L

Lamp Module 38–40
　　installing 39–40

lights, controlling 37–45
　　applications 45
　　overview 37

M

Macintosh 93–132
　　changing module information 117–118
　　changing the background 123–126
　　checking the Events Meter 116
　　creating and using module icons 100–106
　　deleting module icons 121
　　getting started 96–100
　　Instant X-10 desk accessory 129–131
　　overview 93
　　printing 126–127
　　removing timed events 113
　　reviewing timed events 114–115
　　saving files 127
　　setting the Base Housecode 118–120
　　setting the interface clock 121–122
　　setting up the interface 94–95
　　setting up timed events 106–113
　　testing the interface 99–100

Maxi Controller 10–13
　　comparison with other controllers 13
　　overview 10
　　using 10–13

Mini Controller 14–16
　　comparison with other controllers 16
　　overview 14
　　using 14–16

Mini Timer 17–26
　　comparison with other controllers 26
　　installing the backup battery 25
　　overview 17
　　using for manual control 18
　　using for timed events 18–25
　　　　clearing events 24–25
　　　　one-time timed events 22
　　　　overview 18
　　　　reviewing timed events 24
　　　　Security mode 23
　　　　setting the clock 18–19
　　　　setting the Wake Up Alarm 21–22
　　　　setting timed events 19–21
　　　　Sleep mode 23

modules 2–3
　　setting addresses 4–6

P

PL513 transmitter 162

Power Line Carrier 157

Powerflash Burglar Alarm Interface 70–74

problems, troubleshooting 163–165

R

Remote Control and Wireless Transceiver 33–36
　　comparison with other controllers 36
　　overview 33–35

rocker buttons 14, 15

S

security *see* home security

sprinklers, controlling 61–62

Supervised Home Security System *see* home security

T

technical overview 157–162

techno-speak 31

Telephone Transponder 28–33

    comparison with other controllers 33
    overview 28–29
    using for manual control 29–30
    using for remote control 30–33
        remote operation 32
        setting the ANSWERING MACHINE switch 32
        setting the security code 30–31
        setting up the transponder 31

Thermostat Set-back Controller 52–55

Three-Way Wall Switch Module 42–45
    installing 43–45

timed events 17 *see also* IBM PC, Macintosh

troubleshooting problems 163–166

TW523 transmitter/receiver 162

Two-Way Wall Switch Module 40–42
    installing 41–42

U

Unit Code buttons 10

Unit Codes 5

Universal Module 57–65

W

Wall Receptacle Module 50–52

Wireless Motion Detector 86–89

Wireless Transceiver 33–35

X

X-10–compatible products 7, 167–173

Z

zero crossing point 157